the new
science
of
management
decision

revised edition

the new science of management decision

Herbert A. Simon

Professor of Computer Science and Psychology
Carnegie-Mellon University

PRENTICE-HALL, INC., Englewood Cliffs, New Jersey 07632

Library of Congress Cataloging in Publication Data

SIMON, HERBERT ALEXANDER, (date)
  The new science of management decision.

  Edition of 1965 published under title: The shape of
automation for men and management.
  Includes bibliographical references and index.
  1. Decision-making—Data processing.   2. Industrial
management—Data processing.   3. Automation—Economic
aspects.   I. Title.
HD69.D4S49 1977   658.4′03   76-40414
ISBN 0-13-616144-8
ISBN 0-13-616136-7 pbk.

Earlier editions published by Harper & Row, Publishers

Printed in the United States of America

10   9   8   7   6   5   4   3   2   1

Prentice-Hall International, Inc., *London*
Prentice-Hall of Australia Pty. Limited, *Sydney*
Prentice-Hall of Canada, Ltd., *Toronto*
Prentice-Hall of India Private Limited, *New Delhi*
Prentice-Hall of Japan, Inc., *Tokyo*
Prentice-Hall of Southeast Asia Pte. Ltd., *Singapore*
Whitehall Books Limited, *Wellington, New Zealand*

To Katherine, Pete, Barbara,
and their fellow trustees,

Who hold the world in trust for
Peter, Andrew, Matthew and Rachel
and so many others.

contents

4

Organizational Design: Man-Machine Systems

5

preface
to the
revised edition

This is really the third edition of a volume that first appeared in 1960, in which I have brought together some of my thinking on the subject of computers and their implications for business management. The first edition was based on a series of lectures delivered at the School of Commerce, Accounts, and Finance, New York University, during my tenure there as Ford Distinguished Visiting Professor. Five years later, I added to the text an essay on the economic effects of automation, together with other material, and brought out a second edition under the title of *The Shape of Automation*. (I apologize to librarians and others for the confusion caused by this change in title, and return with this new edition to the original title, which seems to me to describe better than the other the main contents and emphases of the book.) My motives in writing the book, and in now substantially revising and enlarging it, can still be stated in the words of the preface to the first edition:

> My research activities during the past decade have brought me into contact with developments in the use of electronic digital computers. These computers are startling even in a world that takes atomic energy and the prospects of space travel in its stride. The computer and the new decision-making techniques associated with it are bringing changes in white-collar, executive, and professional work as momentous as those the introduction of machinery has brought to manual jobs. These essays record the product of my reflections about the organizational and social implications of these rapid technical developments.
>
> I do not apologize for extrapolating beyond our present certain knowledge. In our kind of world, those who are closest to important new technical innovations have a responsibility to provide reasoned interpretations of these innovations and their significance. Such interpretations should be, of course, the beginning and not the end of public discussion. But they cannot be made at all without extrapolation from present certainties into future probabilities.

Now ten years have passed since the appearance of the second edition and over fifteen since the appearance of the first. It is time to reassess conclusions and trends, to revise them where necessary, and to project them again into the future. For this third edition, the text has been expanded and completely revised, particularly in its discussions of the human consequences of automation and the implications of computers for top management and for organization structure. I will not say that its conclusions have been changed dramatically, for the original prophecies have mostly proved quite correct. But now these same conclusions, prospective in 1960, can be buttressed by factual evidence, and can be modified where they have been shown to be inaccurate. In 1960 we stood at the dawn of the computer era in management; today we are well into its morning. In the course of revising the manuscript I found myself changing many sentences from the future tense to the present.

Chapter I of the present edition is reprinted by permission and with considerable abridgement and revision, from *Manage-*

ment and Corporations, 1985.[1] I wish to thank New York University for relinquishing the rights to such of the remaining materials as are based on previous editions.

I have acknowledged at several points in the text my debt to the colleagues who have been my partners in research in this area —and particularly to Allen Newell. My debts go far beyond these specific acknowledgements—how far I cannot specify without implicating my friends in conclusions with which they might wish to disagree. I am especially grateful to my wife, Dorothea Simon, who undertook much of the literature search for this third edition, and who has contributed numerous improvements to successive drafts of the manuscripts of all three editions and participated in their editing.

Herbert A. Simon
Pittsburgh, Pennsylvania

[1] M. Anshen and G. L. Bach, eds. (New York: McGraw-Hill Book Company, 1960).

**the new
science
of
management
decision**

introduction:
computers
and
management

The main purpose of this book is to examine how the processes of management, and especially management decision making, have changed and continue to change under the impact of the new technology of the computer. To understand the nature and import of these changes they must be viewed in a wider context—the whole context of the mechanization and automation of work and the workplace, and of the continuing advance of technology. For that reason, we will need to concern ourselves with the wider social effects of computers as well as their immediate consequences for the processes of making decisions.

Computers and automation have captured man's imagination.[1] That is to say, like the psychiatrist's ink blot, they serve the

[1] For simplicity, the words "man", "he," and "his" are sometimes used generically in this book to denote "person."

1

imagination as symbols for all that is mysterious, potential, portentous. For when man is faced with ambiguity, with complex shadows he only partly understands, he rejects that ambiguity and reads meanings into the shadows. And when he lacks the knowledge and technical means to find the real meanings of the shadows, he reads into them the meanings in his own heart and mind, uses them to give external shape to his private hopes and fears. So the ambiguous stimulus, the ink blot, becomes a mirror. When man describes it, he depicts not some external reality, but himself.

Computers are splendid ink blots. The rash of computer cartoons that decorate the pages of popular magazines gives evidence of their reflecting power. The cartoon computer is Cornucopia, promising plenty without toil. It is a tireless worker, displacing man from his job, preventing him from sharing in plenty. It is a Golem, disguising itself as a man, conquering Man, subjecting him to the rule of Machine. (But the cartoon computer has its human failings too: it plays practical jokes, it makes Gargantuan mistakes.) So the cartoon computer serves as a mirror of our hopes and anxieties. We hope for a world free from poverty and excessive toil. We worry lest our role in society be abolished by social change. We prize our human uniqueness, and are anxious for the safety of our human freedoms.

All of these desires, concerns, and fears are understandable and laudable. Creatures insensitive to their own preservation do not survive. But as long as the ink blot is inscrutable, it does not tell us whether these hopes are realistic, whether the fears are well founded. Worse, since we do not really see the ink blot, only our reflection in it, we do not decipher it, we do not discover how to realize the hopes or to avert the dangers that we fear. Perhaps we are not even looking at the right ink blot. Perhaps we should fix our hopes for plenty on outer space rather than on automation; perhaps, as many think today, the atom or the potential destruction of our environment are more deserving of our fear than is the computer. Only by resolving the ambiguity of the computer, only by tracing through and understanding the implications of automation, can we decide.

There are hundreds of thousands of people in the world today who know about computers—who understand the physical principles that enable them to do arithmetic and manipulate symbols, or who can program them, maintain and repair them, even design them. There are millions of persons employed in automated and semi-automated factories and offices who have daily contact with computers or the information produced by computers. This does not mean at all that we "understand" computers—in particular, that we understand the potential range of their capabilities and applications, or the consequences they will have for business and for society. We only have to look back at the history of the automobile, the airplane, or the radio to see how different is the understanding required to introduce and develop a new piece of technology from the understanding needed to assess its economic and social implications, to anticipate them, and to deal with them.

It should be no surprise, then, that in the field of computers and automation even the experts disagree violently—to say nothing of those who comment on these matters without being expert. (Let me arbitrarily define an *expert* as someone who has participated extensively enough in the development and introduction of computers, computer programing, and automation to have a broad grasp of the technology.) On the Left—to use the conventional order of seating—we have the late Norbert Wiener, concluding his final book [2] with the sentence, "Since I have insisted upon discussing creative activity under one heading, and not parceling it out into separate pieces belonging to God, to man, and to the machine, I do not consider that I have taken more than an author's normal liberty in calling this book *God and Golem, Inc.*" On the Right, we have the late Mortimer Taube, a competent engineer in the field of information retrieval, leveling charges of stupidity, if not fraud, against those who would juxtapose the words *thinking* and *machine* [3]; and, more recently, Joseph Weizenbaum, who inveighs against the potential harm and present immorality of research in artificial intelligence.[4]

Much of the writing on these topics is passionate—not only

[2] *God and Golem, Inc.* (Cambridge, Mass.: M.I.T. Press, 1964).

[3] *Computers and Common Sense* (New York: Columbia Univ. Press, 1961).

[4] *Computer Power and Human Reason* (San Francisco: W. H. Freeman, 1976). My analysis in the present book may help show why Weizenbaum's worries are misdirected.

the writings of experts, but that of such laymen as the philosopher Hubert Dreyfus, who has discussed at length *What Computer's Can't Do.*[5] The air is filled with accusations, charges, and sometimes countercharges. To many who have engaged in the discussion the inkblot is frightening, threatening. I do not think that the passion and invective have illuminated the important issues, and I will try to resist the temptation to reply in kind. This book undertakes an analysis of the impact of modern management science and computers upon management decision. It is not a defense of the computer.

The disagreement among experts cannot be described, however, along a single continuum of Left and Right. There is a technological dimension, a socioeconomic dimension, and a philosophical dimension, and beliefs along the socioeconomic dimension are almost independent of beliefs along the others—thus creating the potential for at least four parties. Let me characterize the poles on each dimension.

The Technological Dimension

This dimension relates to the technical capabilities, present and potential, of computers and their applications to automation. The key question is: What can a computer (now or in the future) be designed and built to do?

technological radicalism

The radical expert's answer to this question is: "Computers will be able to do any work a man can do." Of course, there are degrees of radicalism, some experts looking to a distant future when the prediction will be true, others adding the phrase, "in our time."

technological conservatism

The conservative states, "Computers can do only what you program them to do," and then derives all sorts of conclusions about what computers can never do from this unimpeachable premise.

[5] New York: Harper & Row, 1972.

The extreme conservative asserts that computers cannot do anything they are not already doing—a claim that, in a way, never needs to be revised, no matter how rapidly the world changes.

The Economic Dimension

This dimension relates to the impact of computers and automation on the economy: the rate at which they will be introduced, the consequences for productivity and employment, and the effects upon the structure and operation of business and governmental organizations.

economic radicalism

The (computer) radical predicts an enormous flow, or "glut," of goods and services, mass unemployment, particularly among the unskilled, and organizations in which the few remaining men either are the system's masterminds, or, alternatively, are held in the cold embrace of the machine.

economic conservatism

The economic conservative sees automation as simply the continuation of the industrial revolution. Productivity will continue to rise, not much more rapidly than before, availability of capital will limit the rate of rise, the economy will maintain substantially full employment, and the economic world will continue to look much as it has in the past.

The Philosophic Dimension

This dimension is not entirely distinct from the technological, but is less concerned with the potentials of computers for the world of work than with their implications for the uniqueness of Man. Does a human being have capabilities that, intrinsically and essentially, are beyond those of present or future computers?

philosophic radicalism

The philosophic radical, like the technological radical, believes that, in principle, the cognitive abilities of computers are coextensive with the abilities of Man. He may or may not believe that the full scope of these abilities can be realized with the hardware configurations we have now—some think that computer memories will have to grow much larger to match human memories, others that computers will need more capabilities for parallel operation to equal the real-time speed of human thinking and perception. The radical may become more cautious when asked whether computers can have consciousness or emotion, and may restrict his claims to the cognitive domain. At most, he is likely to answer that these matters are not clear, and are unlikely to become clear until the terms of the questions are defined more sharply.

philosophic conservatism

The philosophic conservative draws a sharp and immovable boundary between man and machine. He uses a variety of arguments to locate this boundary: for example, the hypothetical exemption of man, but not machine, from the Gödel theorems of modern logic, which demonstrate that certain true statements are eternally unprovable by machine; the asserted capability of humans to perceive instantaneously the holistic, Gestalt properties of complex stimuli; or the supposedly unique ability of humans to make value judgments.

I might as well state at the outset my own coordinates along these three dimensions: the technological, the economic, and the philosophic. Since my position is fairly extreme along all dimensions, it is not too hard to characterize: I am a technological radical, an economic conservative, and a philosophic pragmatist. Stripping away minor qualifications and reservations, which will be supplied later, I believe that in our time computers will be able to perform any cognitive task that a person can perform. I believe that computers already can read, think, learn, create. I believe that computers and automation will contribute to a continuing, but not greatly accelerated, rise in productivity, that full employment will be maintained in the face of that rise, and that mankind will not find the life of production and consumption in a more automated world greatly different from what it has been in the past.

On the philosophic side, I find myself unconvinced by the arguments that purport to show there are limits on machines that do not apply to man, but I am willing to hold judgment in abeyance as we explore the capabilities of computers more fully. If there are limits to that exploration, we will sooner or later discover them, and then we will also understand the philosophic arguments better. The failure of philosophic attempts in past ages to prove, from first principles generated in an armchair, what is possible or impossible do not generate confidence in the similar arguments now applied to the capabilities of computers.

Which Expert?

How does one choose among experts? The easiest and commonest way is to accept an expert who confirms one's present beliefs and prior prejudices. In the area of computers and automation, there is little difficulty in finding such an expert, no matter what the beliefs or the prejudices. The hardest way to choose among experts is to decide which expert is right. To do this without first becoming an expert yourself is difficult; nevertheless, it is not impossible—our faith in democracy is an affirmation of that possibility.

We choose among experts by forcing the experts to disclose how they reached their conclusions, what reasoning they employed, what evidence they relied upon. Having made this disclosure, they have exposed their assumptions and reasoning to the scrutiny and challenge of other experts. We do not have to be championship boxers to referee a fight.

But disclosure serves a second purpose. The expert is seldom an expert about consequences and policy implications: he may not even be much of an expert about parts of the technology that are distant from his particular professional concerns. When the expert explains how he has reached his conclusions, we generally find that much of his reasoning derives from the same common sense and common knowledge of the world that all of us believe ourselves to possess. When he claims his expertness as the direct source of his conclusions, we are reluctant to challenge them. But we are quite prepared to examine and judge for ourselves those tenuous paths of common-sense dialectic that commonly connect a specialized fact with a general consequence in the real world. We can do this if the paths are revealed.

Organization of the Essays

In this book, therefore, I have undertaken not only to describe the processes of management decision making and the impact of computers on those processes and on the larger systems in which they take place, but also to state how I have reached my conclusions about them. The first chapter gives an introductory view of the new computer technology and its implications for society, for organizations, and for management. It touches upon most of the topics that are developed in later chapters, leaving to them a more detailed examination of the evidence for its conclusions.

The second chapter analyzes the management decision process and undertakes to provide a nontechnical description of what computers can now do, and soon will be able to do, to contribute to that process. The third chapter discusses the impact of the computer and automation on the workplace, their effects upon job satisfaction and upon worker motivations and alienation. The fourth chapter explores the ways in which computers are changing and may change the job of the manager and the structure of the organization. The fifth chapter returns to a view of the broader social scene, examining the economic and social effects of automation, and of technological advance generally.

Nothing has happened in the past fifteen years that has changed in essential ways my assessment of the development of computer technology or its implications for organization and management. The state of the art has progressed rapidly, as was predictable and predicted, but with minor qualifications it has not progressed in unexpected ways. Hence revising this volume, although the revision is extensive, has been more a matter of updating the evidence and treating a number of topics more fully, than it has been a matter of altering the main findings and conclusions.

In the computer field, the moment of truth is a running program; all else is prophecy. In the chapters that deal with the technology, I have tried to give a general description of some of the computer programs operating today on which my conclusions and forecasts rest. A reader who wants a more up-to-date review of *avant-garde* programs will find numerous examples in *Semantic Information Processing*,[6] and *Representation and Meaning*.[7] A brief introduction to modern information processing theories of human thinking can be found in my book, *The Sciences of the Artificial*,[8] and a more complete treatment in Allen Newell and Herbert A. Simon, *Human Problem Solving*.[9]

[6] Marvin Minsky, ed. (Cambridge, Mass.: MIT Press, 1968).
[7] L. Siklóssy and H. A. Simon, eds. (Englewood Cliffs, N.J.: Prentice-Hall, 1972).
[8] H. A. Simon (Cambridge, Mass.: MIT Press, 1969).
[9] A. Newell and H. A. Simon (Englewood Cliffs, N.J.: Prentice-Hall, 1972).

1

will the corporation
be managed
by machines?

    Some years ago a colleague assigned the title of this chapter to a talk I had agreed to give.[1] Not knowing whether it was assigned seriously or humorously, I took it seriously. I had been too close to computers for too long to treat it lightly. Perhaps I had lost my sense of humor and perspective about them.

    My work on the talk was somewhat impeded by a fascinating spectacle going on just outside my office window. Men and machines were constructing the foundations of a small building. After some preliminary skirmishing by men equipped with

---

[1] In preparing the initial version of this chapter, I drew heavily on two previous essays written in collaboration with Allen Newell: "Heuristic Problem Solving: The Next Advance in Operations Research," *Operations Research,* 6, January-February 1958, 1-10; and "What Have Computers to Do With Management?", C. P. Shultz and T. L. Whisler, eds., *Management Organization and the Computer* (New York: The Free Press, 1950).

surveying instruments and sledges for driving pegs, most of the work was done by various species of mechanical elephants and their mahouts. Two kinds of elephant dug out the earth (one with its forelegs, the other with its trunk) and loaded it in trucks (pack elephants, I suppose). Then, after an interlude during which another group of men carefully fitted some boards into place as forms, a new kind of elephant appeared, its belly full of concrete which it disgorged into the forms. It was assisted by two men with wheelbarrows—plain old-fashioned man-handled wheelbarrows—and two or three other men who fussily tamped the poured concrete with metal rods. Twice during this whole period a shovel appeared—on one occasion it was used by a man to remove dirt that had been dropped on a sidewalk; on another occasion it was used to clean a trough down which the concrete slid.

Here, before me, was a sample of automated, or semiautomated production. What did it show about the nature of present and future relations of man with machine in the production of goods and services? And what lessons that could be learned from the automation of manufacturing and construction could be transferred to the problems of managerial automation? I concluded that there were two good reasons for beginning my analysis of the use of computers by management with a careful look at factory and office automation. First, the business organization is becoming and will increasingly become a highly automated man-machine system, and the nature of management will surely be conditioned by the character of the system being managed. Second, perhaps there are greater similarities than appear at first blush among the several areas of potential automation—blue collar, clerical, and managerial. Perhaps the automated executive of the future has a great deal in common with the automated worker or clerk whom we can already observe in many situations today.

First, however, we must establish a framework and a point of view. Our task is to examine the changing job of the manager, and to forecast the continuing alterations that will take place in that job as computers come to play a larger role in the decision-making process. It is fair to ask: Which manager? Not everyone nor every job will be affected in the same way; indeed, most persons who will be affected are not even managers at the present time. Moreover, we must distinguish the gross effects of a technological change, which occurs at the point of impact of that change, from the net effects—the whole series of secondary ripples spreading from that point of initial impact.

Many of the initial effects are transitory—important enough to those directly involved at the time and place of change, but of no lasting significance to the society. Other effects are neither apparent nor anticipated when the initial change takes place, but flow from it over a period of years through the succession of reactions it produces. Examples of both transient and indirect effects of change come to mind readily enough—e.g., the unemployment of blacksmiths and the appearance of suburbs, respectively, as effects of the growing use of the automobile.

In this book I shall say little about the transient effects of the change in the job of the manager or of workers. I do not mean to discount the importance of these effects to the people they touch. In our time we are highly conscious of such effects, particularly the harmful ones—the displacements of skill and status. We say less of the benefit to those who acquire the new skills or of the exhilaration that many derive from erecting new structures.

Of course, the social management of change does not consist simply in balancing beneficial temporary consequences against harmful ones. The simplest moral reasoning leads to a general rule for the introduction of change: The general society which stands to benefit from the change should pay the major cost of introducing it and should compensate generously those who would otherwise be harmed by it. A discussion of the transient effects of change would have to center on ways of applying that rule. But that is not the problem we have to deal with here.

Our task is to forecast the longer-run effects of change. First of all, we must predict what is likely to happen to the job of the individual manager and to the activity of management in the individual organization. Changes in these patterns will have secondary effects on the occupational profile in the economy as a whole. Our task is to picture the society after it has made all these secondary adjustments and settled down to its new equilibrium.

Let me now indicate the general plan I shall follow in this chapter. In the first section, "Predicting Long-Run Equilibrium," I shall identify the key factors—the causes and the conditions of change—that will mold the analysis. Then I shall show how a well-known tool of economic analysis—the doctrine of comparative advantage—permits us to draw valid inferences from these causes and conditions. The economic argument will be stated quite briefly in this section. Readers who are skeptical of the conclusions, or who would like to pursue the issue in greater depth, will find a fuller treatment of it in Chapter 5.

In the second section, "The New Technology of Information Processing," I shall describe some of the technological innovations that have appeared and are about to appear in the areas of production and data processing, and I shall use this material to draw a picture of the emerging business organization, with particular attention to the automation of blue-collar and clerical work. Again, these topics will be developed in greater detail in Chapters 2 and 3.

In the third section, "The Automation of Management," I shall consider more specifically the role of the manager in the future business organization. This topic, in turn, is expanded in Chapter 4, below. Finally, in the section entitled, "The Broader Significance of Automation," I shall try to identify some of the important implications of these developments for our society and for ourselves as members of it. These implications will be discussed further in Chapter 5.

Predicting Long-Run Equilibrium

To predict long-run equilibrium, one must identify two major aspects of the total situation: (1) the variables that will change autonomously and inexorably—the "first causes," and (2) the constant, unchanging "givens" in the situation, to which the other variables must adjust themselves. These are the hammer and the anvil that beat out the shape of the future. The accuracy of our predictions will depend less upon forecasting exactly the course of change than upon assessing correctly which factors are the unmoved movers and which the equally unmoved invariants. My entire forecast rests on my identification of this hammer and this anvil.

the causes of change

The growth in human knowledge is the primary factor that will give the system its direction—in particular, that will fix the boundaries of the technologically feasible. The growth in real capital is the major secondary factor in change—within the realm of what is technologically feasible, it will determine what is economical.

The crucial area of expansion of knowledge is not hard to predict, for the basic innovations—or at least a large part of them —have already occurred, and we are now rapidly exploiting them. The new knowledge consists in a fundamental understanding of the processes of thinking and learning, or to use a more neutral term, of complex information processing. We can now write programs for electronic computers that enable these devices to think and learn.[2] This knowledge is having, and will have, practical impacts in two directions: (1) because we can now simulate in considerable detail an important and increasing part of the processes of the human mind, we have available a technique of tremendous power for psychological research; (2) because we can now write complex information-processing programs for computers, we are acquiring the technical capacity to replace humans with computers in a rapidly widening range of "thinking" and "deciding" tasks. It is important that we explore in both of these directions, and not neglect one in favor of the other. We are talking about the direct enhancement of human intelligence by gaining a deeper understanding of how the human mind works; and we are talking about the indirect enhancement of human intelligence by augmenting it with the artificial intelligence of computers.

Closely allied to the development of complex information-processing techniques for general-purpose computers is the rapid advance in the technique of automating all sorts of production

[2] For documentation of this claim, see Chapter 2 under "Heuristic Problem Solving," p. 62 ff.

and clerical tasks. Putting these two lines of development together, I am led to the following general predictions: Within the near future—perhaps in the next generation—we shall have the *technical* capability of substituting machines for any and all human functions in organizations. Within the same period, we shall have acquired an extensive and empirically tested theory of human cognitive processes and their interaction with human emotions, attitudes, and values.

To predict that we will have these technical capabilities says nothing of how we shall use them. Before we can forecast that, we must discuss the important invariants in the social system.

the invariants

The changes that our new technical capability will bring about will be governed, particularly in the production sphere, by two major fixed factors in the society. Both of these have to do with the use of human resources for production.

1. Apart from transient effects of automation, the human resources of the society will be substantially fully employed. Full employment does not necessarily mean a forty-hour week, for the allocation of productive capacity between additional goods and services and additional leisure may continue to change as it has in the past. Full employment means that the opportunity to work will be available to the great majority of adults in the society and that, through wages and other allocative devices, the product of the economy will be distributed widely among families.

Since, at the very time when I am writing these lines, unemployment in the United States hovers around 8 or 9 percent of the labor force, it may seem strange to make so causally the assumption of full employment. Nevertheless, our society is committed to a policy of full employment, and as our understanding of economic mechanisms has grown over the past half century, our ability, through government fiscal and monetary policy, to moderate the severity and duration of periodic depressions has grown with it. We can realistically speak of employment for most people most of the time. Most important, we have learned that increased productivity in general and increased mechanization or automation in particular do not carry with them permanent unemployment as a consequence. Two centuries of technological progress have demonstrated that any level of employment is compatible with any level of productivity.[3]

2. The distribution of intelligence and ability in the society will be much as it is now, although a substantially larger percentage of adults (perhaps half or more) will have completed college educations.

These assumptions—of the capability for automation accompanied by full employment, and constancy in the quality of the human resources—provide us with a basis for characterizing the changes produced and to be produced by automation. We cannot talk about the technological unemployment it may create, for we have concluded that such unemployment is a transient phenomenon—that there will be none in the long run. But the pattern of occupations, the profile showing the relative distribution of employed persons among occupations, may be greatly changed. It is the change in this profile that will measure the organizational impact of the technological change.

The change in occupational profile depends on a well-known economic principle, the doctrine of comparative advantage. It may seem paradoxical to think that we can increase the productivity of mechanized techniques in all processes without displacing men somewhere. Won't a point be reached where people are less productive than machines in *all* processes, hence economically unemployable?

---

[3] In previous depressions, for example in the 1930s and in 1965, unemployment was frequently attributed by public opinion to technological progress in general, and mechanization in particular. President Johnson felt obliged to appoint a Presidential Commission, the National Commission on Technology, Automation and Economic Progress, to examine the validity of this charge. The Commission's report, published by the U. S. Government Printing Office, 1966, as *Technology and the American Economy* reached the same conclusion as the one I have just stated—that there is no logical connection between mechanization and long-term unemployment. It is interesting that automation has hardly been mentioned during the current depression, in spite of its severity. Perhaps the economics lesson has been learned, or perhaps attention was diverted to other putative causes by the environmental and energy crises.

The paradox is dissolved by supplying a missing term. Whether man or machines will be employed in a particular process depends not simply on their relative productivity in physical terms but on their cost as well. And cost depends on price. Hence—so goes the traditional argument of economics— as technology changes and machines become more productive, the prices of labor and capital will so adjust themselves as to clear the market of both. As much of each will be employed as offers itself at the market price, and the market price will be proportional to the marginal productivity of that factor. By the operation of the market place, manpower will flow to those processes in which its productivity is comparatively high relative to the productivity of machines; it will leave those processes in which its productivity is comparatively low. The comparison is not with the productivities of the past, but among the productivities in different processes with the currently available technology.

I apologize for dwelling at length on a point that is clearly enough stated in *The Wealth of Nations.* My excuse is that contemporary discussion of technological change and automation still very often falls into error through not applying the doctrine of comparative advantage correctly and consistently.

We conclude that human employment will become smaller relative to the total labor force in those kinds of occupations and activities in which automatic devices have the greatest comparative advantage over humans; human employment will become relatively greater in those occupations and activities in which automatic devices have the least comparative advantage.[4]

---

[4] I am oversimplifying, for there is another term in this equation. With a general rise in productivity and with shifts in relative prices due to uneven technological progress in different spheres, the demands for some kinds of goods and services will rise more rapidly than the demands for others. Hence, other things being equal, the total demand will rise in those occupations (of men and machines) that are largely concerned with producing the former more rapidly than in occupations concerned largely with producing the latter. I have shown elsewhere how all these mechanisms can be handled formally in analyzing technological change. See "Productivity and the Urban-Rural Population Balance," *Models of Man* (New York: John Wiley & Sons, Inc., 1957), chap. 12; and "Effects of Technological Change in a Linear Model," T. Koopmans, ed., *Activity Analysis of Production and Allocation* (New York: John Wiley & Sons, Inc., 1951), chap. 15; see also pp. 21-27 below.

Thus, if computers are a thousand times faster than bookkeepers in doing arithmetic, but only one hundred times faster than stenographers in taking dictation, we shall expect the number of bookkeepers per thousand employees to decrease but the number of stenographers to increase. Similarly, if computers are a hundred times faster than executives in making investment decisions, but only ten times faster in handling employee grievances (the quality of the decisions being held constant), then computers will be employed in making investment decisions, while executives will be employed in handling grievances.

The New Technology of Information Processing

The automation of manufacturing processes is a natural continuation and extension of the Industrial Revolution. We have seen a steady increase in the amount of machinery employed per worker. In the earlier phases of mechanization, the primary function of machinery was to replace human energy with mechanical energy. To some extent in all phases, and to a growing extent in recent developments, another goal has been to substitute mechanical for human sensing and controlling activities. Those who distinguish the newer "automation" from the older "mechanization" stress our growing ability to replace with machines simple human perceiving, choosing, and manipulating processes.

the nearly automatic factory

The genuinely automatic factory—the workerless factory that can produce output and perhaps also, within limits, maintain and repair itself—will likely be technically feasible, at least in some process industries, within a generation. From very unsystematic observation of changes going on in factories today, one might surmise that the typical factory will not, however, be fully automatic for some time to come. More likely the typical factory will have reached, say, the level of automaticity that has now been attained by the most modern oil refineries or power generating stations.

The same kinds of technical developments that lead toward the automatic factory have been bringing about an even more rapid revolution—and perhaps eventually a more complete one —in large-scale clerical operations. The abstract nature of symbol manipulation facilitates the design of equipment to do it, and the further automation of clerical work is impeded by fewer technical barriers than the further automation of factory production. The departments of a company concerned with major clerical functions—accounting, processing of customers' orders, inventory and production control, purchasing, and the like—are already reaching an even higher level of automation than most factories.

Both the factory and the office, then, are rapidly becoming complex man-machine systems with a very large amount of production equipment, in the case of the factory, and computing equipment, in the case of the office, per employee. The clerical department and the factory will come more and more to resemble each other. The one will present the picture of a small group of employees operating (I am tempted to use the more accurate phrase, "collaborating with") a large data-processing system; the other, the picture of a similar small group of employees operating a large automated production system. The interrelation of man with machine will become—is already becoming—quite as important a design problem for such systems as the interrelation of man with man.

Now we must not commit the error I warned against in discussing the doctrine of comparative advantage. When we foresee fewer employees in factory and office, we mean fewer per unit of output and fewer per unit of capital equipment. It does not follow that there will be fewer in total. To predict the occupational profile that will result, we must look more closely at the prospective rates of automation in different occupations.

Before we turn to this task, however, it is worth reporting a couple of the lessons that are currently being learned in factory and clerical automation, the evidence for which will be presented in Chapter 3:

1. Automation does not mean "dehumanizing" work. On the contrary, in most actual instances of recent automation, jobs were made on the whole at least as pleasant and interesting, as judged by the employees themselves, as they had been before.

In particular, automation is moving more and more in the direction of eliminating the machine-paced assembly line task and the repetitive clerical task. It appears generally to reduce the "work-pushing," "man-driving," and "expediting" aspects of first-line supervision.

2. Contemporary automation does not generally change to an important extent the profile of skill levels among the employees. It perhaps calls on the average for some upgrading of skills in the labor force, but as we shall see in Chapter 3, conflicting trends have been observed at different stages in automation.

the occupational profile

To predict the occupational distribution of the employed population in a highly automated economy, we would have to go down the list of occupations and assess for each the potentialities of automation. Even if we could do this, our inferences could not be quite direct. For we also have to take into account (1) income elasticity of demand—the fact that as productivity rises, the demands for some goods and services will rise more rapidly than the demands for others; (2) price elasticity of demand—the fact that the most rapidly automated activities will also show the greatest price reductions, so that the net reduction in employment in these activities will be substantially less than the gross reduction at a constant level of production.

As a fanciful example, let us consider the number of persons engaged in the practice of psychiatry. It is reasonable to assume that the demand for psychiatric services, at constant prices, will increase more than proportionately with an increase in income. Hence, the income effect of the general increase in a society's productivity will be to increase the proportion of psychiatrists in the employed population. Now, let us suppose that a specific technological development permits the automation of psychiatry itself, so that one psychiatrist can do the work formerly done by ten.[5] It is not at all clear whether a 90 percent

---

[5] This example will seem entirely fanciful only to persons not aware of some of the research now going on into the possible automation of psychiatric processes.

reduction in price of psychiatric services would increase the demand for those services by a factor of more or less than ten. But if the demand increased by a factor of more than ten, the proportion of persons employed in psychiatry would actually increase. Thus prediction of the occupational profile depends on estimates of the income and price elasticity of demand for particular goods and services as well as estimates of relative rates of increase in productivity.

This is not the only difficulty the forecaster faces. He must also be extremely cautious in his assumptions as to what is, and what is not, likely to be automated. In particular, automation is not the only way to reduce the cost of a process—a more effective way is to eliminate it. An expert in automation would tell you that the garbage collector's job is an extremely difficult one to automate (at any reasonable cost) in a straightforward way. It has, of course, been partly eliminated in many communities by grinding the garbage and transporting it in liquid through the sewerage system. Such Columbus-egg solutions of the production problem are not at all rare, and will be an important part of automation.[6]

another approach to prediction

With all these reservations and qualifications, is any prediction possible? I think it is, but I think it requires us to go back to some fundamentals. The ordinary classification of occupations is basically an "end-use" classification—it indicates what social function is performed by each occupation. To understand automation, we must begin our classification of human activities at the other end—what basic capacities does the human organism bring to tasks, capacities that are used in different proportions for different tasks?

Viewed as a resource in production, a man is a pair of eyes and ears, a brain, a pair of hands, a pair of legs, and some muscles

---

[6] I advise the reader, before he makes up his mind as to what is feasible and infeasible, likely and unlikely, to try out his imagination on a sample of occupations, e.g., dentist, waitress, bond salesman, chemist, carpenter, college teacher.

for applying force. Automation proceeds in two ways: (1) by providing mechanized means for performing some of the functions formerly performed by a man and (2) by eliminating some of these functions. Moreover, the mechanized means that replace the man can be of a general-purpose character (like the man) or highly specialized.

The steam engine and the electric motor are relatively general-purpose substitutes for muscles. A butter-wrapping machine is a special-purpose substitute for a pair of hands which eliminates some eye-brain activities the human butter-wrapper would require. A feedback system for controlling the temperature of a chemical process is a special-purpose substitute for eyes, brain, and hands. A digital computer employed in preparing a payroll is a relatively general-purpose substitute for eyes, brain, and hands. A modern multi-tool milling machine is a special-purpose device that eliminates many of the positioning (eye-brain-hand) processes that were formerly required in a sequence of machining operations.

The early history of mechanization was characterized by: (1) rapid substitution of mechanical energy for muscles; (2) partial and spotty introduction of special-purpose devices that performed simple, repetitive eye-brain-hand sequences; (3) elimination, by mechanizing transport and by coordinating sequences of operations on a special-purpose basis, of many human eye-brain-hand sequences that had previously been required.

Thus man's comparative advantage in energy production has been greatly reduced in most situations, to the point where he is no longer a significant source of power in our economy. He has been supplanted also in performing many relatively simple and repetitive eye-brain-hand sequences. He has retained his greatest comparative advantage in: (1) the use of his brain as a flexible general-purpose problem-solving device, (2) the flexible use of his sensory organs and hands, and (3) the use of his legs, on rough terrain as well as smooth, to make this general-purpose sensing-thinking-manipulating system available wherever it is needed.

This picture of man's functions in a man-machine system was vividly illustrated by the construction work I saw going on outside my window. Most of the energy for earth-digging was being supplied by the mechanical elephants, but each depended on its mahout for eyes and (if you don't object to my fancy) for eye-trunk coordination. The fact that the elephant was operating in rough, natural terrain made automation of the mahout a difficult, although by no means insoluble, technical problem (recall the LEM ambulating on the surface of the moon). It would almost certainly not now be economical. But other men—the men with wheelbarrows particularly—were performing even more "manual" and "primitive" tasks. Again, the delivery of the concrete to the forms would today be more fully automated. However, the men provided a flexible, if not very powerful, means for delivering small quantities of concrete to a number of different points over uneven terrain.

"Flexibility" and general-purpose applicability are the key to most spheres where the human has a comparative advantage over the machine. This raises two questions:

1. What are the prospects for matching human flexibility with automatic devices?

2. What are the prospects for matching human skills in particular activities by reducing the need for flexibility?

The second question is a familiar one throughout the history of mechanization; the first alternative is more novel.

flexibility in automata

We must consider separately the sensory organs, the manipulatory organs, the locomotive organs, and the central nervous system. The problem-solving and information-handling capabilities of the brain have proved to be the easiest to duplicate, and great progress has been made in this direction. But these capabilities are so much involved in management activity that we shall have to discuss them at length in a later section.

We are much further from replacing the eyes, the hands, and the legs. From an economic as well as a technological standpoint, I would hazard the guess that automation of the functions wholly within the central nervous system will be feasible long before automation of comparably flexible sensory, manipulative, or locomotive functions. I shall state later my reasons for thinking this.

If these conjectures are correct, automation of thinking and symbol-manipulating functions should proceed more rapidly than the automation of the more complex eye-brain-hand sequences. Certainly the developments of the past twenty years support this prediction. But before we grasp the conclusion too firmly, as applying to either the future or the past, we need to remove one assumption.

environmental control a substitute for flexibility

If we want an organism or mechanism to behave effectively in a complex and changing environment, we can design into it adaptive mechanisms that allow it to respond flexibly to the demands the environment places on it. Alternatively, we can try to simplify and stabilize the environment. We can adapt organism to environment or environment to organism.

Both processes have been significant in biological evolution. The development of the multicellular organism may be interpreted as simplifying and stabilizing the environment of the internal cells by insulating them from the complex and variable external environment in which the entire organism exists. This is the significance of homeostasis in evolution—that in a very real sense it adapts the environment to the organism (or the elementary parts of the organism) and hence avoids the necessity of complicating the individual parts of the organism.

Homeostatic control of the environment (the environment, that is, of the individual worker or the individual machine) has played a tremendous role in the history of mechanization and in the history of occupational specialization as well. Let me cite some examples that show how all-pervasive this principle is:

1. The smooth road provides a constant environment for the vehicle—eliminating the advantages of flexible legs.

2. The first step in every major manufacturing sequence

(steel, textiles, wood products) reduces a highly variable natural substance (metallic ore, fiber, trees) to a far more homogeneous and constant material (pig iron, thread, boards, or pulp). All subsequent manufacturing processes are thus insulated from the variability of the natural material. The application of the principle of interchangeable parts performs precisely the same function for subsequent manufacturing steps.

3.   By means of transfer machines, work in process in modern automated lines is presented to successive machine tools in proper position to be grasped and worked, eliminating the sensory and manipulative functions of workers who formerly loaded such tools by hand.

We see that mechanization has more often proceeded by eliminating the need for human flexibility—replacing rough terrain with a smooth environment—than by imitating it. Now homeostatic control of the environment tends to be a cumulative process. When we have mechanized one part of a manufacturing sequence, the regularity and predictiveness secured from this mechanization generally facilitates the mechanization of the next stage.

Let us apply this idea to the newly mechanized data-processing area. One of the functions that machines perform expensively and badly at present, and humans rather well, is reading printed text. Because of the variability of type fonts in such text, it would seem that the human eye is likely to retain for some time a distinct comparative advantage in handling it. But the wider the use of machines in data processing, the more pains we will take to prepare the source data in a form that can be read easily by a machine. Thus, if scientific journals are to be read mostly by machines, and only small segments of their scanning presented to the human researchers, we shall not bother to translate manuscripts into linotype molds, molds into slugs, and slugs into patterns of ink on paper. We shall (and are already beginning to) use the typewriter to prepare computer input, preferably by direct transmission from computer console to memory, and simply bypass the printed volume or produce it as a last step from computer-readable text stored in the computer memory.[7] Chapter 4 will discuss the important implications of this technical capability for the structure of management information systems.

Now these considerations do not alter our earlier conclusion that humans are likely to retain their comparative advantage in activities that require sensory, manipulative, and motor flexibility (and, to a much lesser extent, problem-solving flexibility). They show, however, that we must be careful not to assume that the particular activities that now call for this flexibility will continue to do so. The stabilization of the environments for productive activity will reduce or eliminate the need for flexible response at many points in the productive process, continuing a trend that is as old as multicellular life. In particular, in the light of what has been said of the feasibility of automating problem solving, we should avoid the simple assumption that the higher-status occupations, and those requiring the most education, are going to be the least automated. There are perhaps as good prospects technically and economically for automating completely the job of a physician (but not a surgeon), a corporate vice-president, or a college teacher as for automating the job of the person who operates a piece of earth-moving equipment.

man as man's environment

In most work situations, an important part of man's environment is man. This is, moreover, an exceedingly "rough" part of his environment. Interacting with his fellow man calls on his greatest flexibility both in sensory activity and response. He must read the nuances of expressions, postures, intonations; he must take into account in numerous ways the individuality of the person opposite him.

[7] The manuscript of this book is being prepared and edited in precisely this way, although since my computer is not connected with the printing presses, it will have to print out human-readable copy that must then again be transformed manually before it can be manufactured into a book. If existing technology were used fully, the entire intermediate manual copying of the manuscript (i.e., by the linotype operator) would be bypassed. Moreover, the book would then be available optionally in man-readable or machine-readable form. Perhaps we will manage that for the fourth edition.

What do we mean by "automating" those activities in organizations that consist in responding to other men? I hardly know how to frame the question, much less to answer it. It is often asserted—even by people who are quite sophisticated on the general subject of automation—that personal services cannot be automated, that a machine cannot acquire a bedside manner or produce the positive affect that is produced by a courteous sales clerk.

Let me, at least for purposes of argument, accept these propositions. (They leave me uneasy, for I am aware of how many people in our own culture have affective relations with such mechanisms as automobiles, rolling mills—and computers.) Accepting them does not settle the question of how much of man's environment in the highly automatized factory or office will be man. For much of the interpersonal activity called for in organizations results from the fact that the basic blue-collar and clerical work has been done by humans, who need supervision and direction. Another large chunk of interpersonal activity is the buying and selling activity—the work of the salesman and the buyer.

As far as supervisory work is concerned, we might suppose that it would decrease in the same proportion as the total number of employees; hence, that automation would not affect the occupational profile in this respect at least. This may be true in first approximation, but it needs qualification. The amounts and types of supervision required by a work force depend on many things, including the extent to which the work pace is determined by the men or by machines and the extent to which the work is prescheduled. Supervision of a machine-paced operation is a very different matter from supervision of an operation where the foreman is required to see that the workers maintain a "normal" pace—with or without incentive schemes. Similarly, a highly scheduled shop leaves room for much less "expediting" activity than one where scheduling is less formal and complete.

As a generalization, we may expect that "work-pushing" and "expediting" will make up a smaller part of the supervisory job in highly automated than in unautomated operations. Whether these activities are replaced, in the total occupational profile, by other managerial activities we shall have to consider a little later.

What about the salesman? I have little basis for conjecture on this point. If we think that buying decisions are not going to be made much more objectively than they have in the past, then we might conclude that the automation of the salesman's role will proceed less rapidly than the automation of many other jobs. If so, selling will account for a larger fraction of total employment.

summary: blue-collar and clerical automation

We can now summarize what we have said about the prospects of the automatic factory and office and about the general characteristics of the organization incorporating them. Clearly, it will be an organization with a much higher ratio of machines to men than is characteristic of organizations today. The people in the system can be expected to play three kinds of roles:

(a) There will be a few vestigial "workmen"—probably a smaller part of the total labor force than today—who will be part of on-line production, primarily doing tasks requiring relatively flexible eye-brain-hand coordination (a few wheelbarrow pushers and a few mahouts). There will be few examples of the assembly line as we know it today.

(b) There will be a substantial number of people whose task is to keep the system operating by preventive and remedial maintenance. Machines will play an increasing role, of course, in maintenance functions, but machine powers are not likely to develop as rapidly, relative to those of men, in this area as on-line activities. Moreover, the total amount of maintenance work—to be shared by men and machines—will increase. For the middle run, at least, this group will probably make up an increasing fraction of the total work force.

(c) There will be a substantial number of people at professional levels, responsible for the design of the product, for the design of the productive process, and for general management. We have still not faced the question of how far automation will go in these areas, and hence we cannot say very firmly whether such occupations will be a larger or smaller part of the whole. Anticipating our later analysis, I will conjecture that they will constitute about the same part as they do now of total factory and office employment.

An increase in the machine-man ratio is not the only change we may expect. In the future production and data-processing organizations, some of the kinds of interpersonal relations—in supervising and expediting—that at present are very stressful for most persons engaged in them, will be substantially reduced in importance.

Finally, in the entire occupied population, a larger fraction of members than at present will be engaged in occupations where "personal service" involving face-to-face human interaction is an important part of the job. It is not easy to forecast what these occupations will be, for the reasons already set forth.

In some respects—especially in terms of what "work" means to those engaged in it—this picture of the automated world of the future does not look drastically different from the world of the present. Under the general assumptions we made—rapid automation, but with full employment and a stable skill profile—it will be a more relaxed place than it is now. As far as man's productive life is concerned, the changes do not appear to be earthshaking.

Confidence in these conclusions can be higher by reason of the fact that the predictions are more than a picture of a distant future. Mainly they are extrapolations of trends that have already been going on for two decades, and about which we now have a great deal of factual evidence, some of which will be reviewed in Chapter 3. Moreover, the conclusions do not depend very sensitively on the exact degree of automation attained, or how soon it occurs: A little more or a little less, a little sooner or a little later, would not change the picture much.

The Automation of Management

I have several times sidestepped the question of how far and how fast we could expect management activities to be automated. I have said something about supervision, but little about the large miscellany of management activities involving decision making, problem solving, and just plain thinking. This topic—the impact of computers on the central management functions—forms the substance of Chapters 2 and 4. I should like to anticipate the conclusions of those chapters, very briefly, in order to round out our present discussion of the automation of industrial and business organizations.

The problems that managers at various levels in organizations face can be classified according to how well structured, how routine, and how cut and dried they are when they arise. On the one end of the continuum are highly programed decisions: routine procurement of office supplies, or pricing standard products; on the other end of the continuum are unprogramed decisions: basic, once-for-all decisions to make a new product line, or strategies for labor negotiations on a new contract. Between these two extremes lie decisions with every possible blend of programed and nonprogramed, routine and nonroutine elements.

There is a rough, but far from perfect, correlation between a manager's organizational level and the extent to which his decisions are programed. On the average, the decisions that the president and vice-president face are less programed than those faced by the factory department head or the factory manager.

decision making by computer

We are now well into a technological revolution of the decision-making process. That revolution has two aspects, one considerably further advanced than the other. The first aspect, concerned largely with decisions close to the programed end of the continuum, is the province of the field called "operations research" or "management science." The second aspect, concerned with unprogramed as well as programed decisions, is the province of a set of techniques that have come to be known as "heuristic programing," or sometimes "artificial intelligence." We are gradually acquiring the technological means, through these techniques, to automate all management decisions, nonprogramed as well as programed. As in the case of nonmanagerial work, however, economic rather than technological factors will determine how rapidly automation proceeds.

Managers are largely concerned with (1) supervising, (2) solving well-structured problems, and (3) solving ill-structured problems. The next chapter will indicate that the automation of the second of these activities—solving well-structured problems—is proceeding extremely rapidly; the automation of the third—solving ill-structured problems—moderately rapidly; and the automation of supervision more slowly. However, we have already concluded that, as less and less work becomes man-paced and more and more of it machine-paced, the nature of supervision will undergo change.

There is no obvious way to assess quantitatively these cross-currents and conflicting trends. We might reasonably conclude that management and other professional activities, taken collectively, may constitute about the same part of the total spectrum of occupations a generation hence as they do now. But there is reason to believe that the kinds of activities that now characterize middle management are being more rapidly and fully automated than the others, and hence will come to have a somewhat smaller part in the whole management picture. These assessments are developed more fully in Chapter 4.

some other dimensions of change in management

There are other dimensions for differentiating management and professional tasks, of course, besides the one we have been using. It is possible that if we described the situation in terms of these other dimensions, the change would appear larger. Let me explore this possibility just a little further.

First, I think we can assert that the manager's time perspective is being gradually lengthened. As automated subsystems take over the minute-by-minute and day-by-day operations of the factory and office, the humans in the system are increasingly occupied with preventive maintenance, with system breakdowns and malfunctions, and—perhaps most important of all—with the design and modification of the systems themselves. The automatic factory increasingly—and subject to all of the qualifications I have introduced—runs itself; the company executives are increasingly concerned not with running today's factory, but with designing tomorrow's. Executives have less and less excuse for letting the emergencies of today steal the time that was allocated to planning for the future. This does not imply that planning is a machineless function—it also is being carried out more and more by man-machine systems, but with perhaps a larger man component and a smaller machine component than day-to-day operations.

Does this mean that executives need a high level of technical competence? Probably not. Most automation calls for increased technical skills for maintenance in the early stages; but the farther automation proceeds, the less those who govern the automated system need to know about the details of its mechanism. The driver of this year's automobile needs to know less about what is under the hood than the driver of a 1910 automobile. The user of this year's computer needs to know less about computer design and operation than the user of a 1950 computer. The manager of a highly automated 1985 factory will need to know less about how things are physically produced in that factory than the manager of a 1960 factory.

Similarly, we can dismiss the notion that computer programers will become a powerful elite in the automated corporation. It is far more likely that the programing occupation will become extinct (through the further development of automatic programing techniques) than that it will become all-powerful. More and more, computers will program themselves; and direction will be given to computers through the mediation of compiling systems that will be completely neutral so far as content of the decision rules is concerned. Moreover, the task of communicating with computers is becoming less and less technical as computers come—by means of compiling techniques—closer and closer to handling the irregularities of natural language.[8]

The Broader Significance of Automation

I have tried to present my reasons for making two predictions that appear, superficially, to be contradictory: that we are

---

[8] We can dismiss in the same way the fears that some have expressed that only mathematicians will be able to cope with a computerized world.

acquiring the technical capability to manage corporations by machine, but that humans, for some time to come, will probably be engaged in roughly the same array of occupations as they are now. I find both predictions reassuring.

Acquiring the technical capacity to automate production as fully as we wish, or as we find economical, means that our per capita capacity to produce will reach a point where no lurking justification will remain for poverty or deprivation. We will have the means to rule out scarcity as mankind's first problem and to attend to other problems that are more serious.

In saying this, I am not unaware of the apparent insatiability of wants. We can, however, make moral distinctions between the neediness of an Indian peasant and the neediness of an American middle-class, one-car family. Nor am I unaware of the environmental pollution or the scarcity of energy that may be created by a thoughtless use of our productive capabilities. Automation is not the cause of these problems, nor would a refusal to proceed with automation alleviate them.[9]

In spite of the continuing trend toward automation, the occupations that humans will find in the corporation of the future will be familiar ones. We can dismiss two common fears: first, the fear of technological unemployment; second, the fear that many people feel at the prospect of fraternizing with robots, or being taken over by them, in an automated world. Fraternize we shall—and already do—but in the friendly, familiar way that we fraternize with our automobiles and our power shovels.

Having dealt with these two issues, we shall be better prepared to face the more fundamental problems of that automated world. These are not new concerns, nor are they less important than the problems of scarcity, of resources, of environmental protection, or of peace. But they are long-range rather than short-range concerns, and hence seldom rise to the head of the agenda as long as there are more pressing issues still around. Three of them in particular, I think, are going to receive increasing atten-

[9] See Chapter 5. See also my essay, "Technology and Environment," in *Management Science*, 19 (June 1973), 1110-21, where I argue that "the solution of today's major social problems will come from more and better technology—not from less technology."

tion as automation proceeds: developing a science of man, finding alternatives for work and production as basic goals for society, and reformulating man's view of his place in the universe.

a science of man

I have stressed the potentialities of the computer and of heuristic programing as substitutes for human work. The research now going on in this area is equally important for understanding how humans perform information-processing tasks—how they think. This research has already made substantial contributions to our understanding of the psychology of cognitive processes, and there are reasons to hope that the potential of the information-processing approach is not limited to cognition but may extend to the affective aspects of behavior as well.

We are making good progress toward constructing psychological theories that are as successful as the theories we have in chemistry and biology today. We are acquiring a pretty good understanding of how the human mind works. That understanding will have obvious and fundamental consequences for both pedagogy and psychiatry. We may expect rapid advances in the effectiveness and efficiency of our techniques of teaching and our techniques for dealing with human maladjustment. We may expect also an increasing ability to train people in effective techniques of problem solving and decision making.

social goals

The continuing rise in productivity is beginning to produce profound changes, different from those caused by the Industrial Revolution, in the role that work plays in man's life and among man's goals. It is hard to believe that in the technologically advanced societies man's appetite for gadgets can continue to expand at the rate required to keep work and production in central roles in the society. And if it did continue to expand, that expansion would soon be halted by resource limits and pollution problems. Even Galbraith's proposal for diverting expenditures from gadgets to social services [10] can only be a temporary expedient. We shall have to, finally, come to grips with the problem of leisure.

Our work-oriented culture still sees a close connection between idleness and deviltry. Our suspicion of leisure is reinforced also by a kind of snobbishness: leisure, we fear, may mean that someone will sit on his back porch and whittle; or go fishing; or drink beer and engage in idle conversation with his friends; or do any one of the multitudinous other tedious or inane things that don't happen to be *our* particular hobbies or vices.

Undoubtedly all of those things will happen. As more leisure is placed at our disposal, we will do many things with it that can hardly be described as "useful," "creative," or even "harmless"— just as we now do in our work. There is no reason to find that alarming. There is great diversity in human tastes for the use of time.

But what of those who have strong needs to succeed at hard tasks? In today's society, organizations satisfy important social and psychological needs in addition to the needs for goods and services. For those who do well in managerial careers, they satisfy needs for success and status. For some executives as well as for others, they are important outlets for creativity. In a society where scarcity of goods and services is of little importance, those institutions whose main function is to deal with scarcity will occupy a less central position than they have in the past. In the future, success in management may carry smaller rewards in prestige and status. Moreover, as the decision-making function becomes more highly automated, corporate decision making may perhaps provide fewer outlets for creative drives than it now does. This last point, however, whether management will be a less creative activity, is debatable, and will be discussed further in the next chapter.

man in the universe

It is only one step from the problem of goals to what psychiatrists now refer to as the "identity crisis," and what used to be called

[10] J. K. Galbraith, *The Affluent Society* (Boston: Houghton Mifflin, 1958).

"cosmology." The developing capacity of computers to simulate man—and thus both to serve as his substitute and to provide a theory of human mental functions—will change man's conception of his own identity as a species.

The definition of man's uniqueness has always formed the kernel of his cosmological and ethical systems. With Copernicus and Galileo, he ceased to be the species located at the center of the universe, attended by sun and stars. With Darwin, he ceased to be the species created and specially endowed by God with soul and reason. With Freud, he ceased to be the species whose behavior was—potentially—governable by rational mind. As we begin to produce mechanisms that think and learn, he has ceased to be the species uniquely capable of complex, intelligent manipulation of his environment.

I am confident that man will, as he has in the past, find a new way of describing his place in the universe—a way that will satisfy his needs for dignity and for purpose. But it will be a way as different from the present one as was the Copernican from the Ptolemaic.

2

the processes
of
management decision

The aim of this chapter is to examine the manager as decision maker. But to understand what is involved in decision making that term has to be interpreted broadly—so broadly as to become almost synonymous with managing.

Our usual image of the decision maker pictures him in a much narrower function. He is the brooding man on horseback who suddenly rouses himself from thought and issues an order to a subordinate. Or he is a happy-go-lucky fellow, a coin poised on his thumbnail, ready to risk his action on the toss. Or she is an alert, gray-haired businesswoman, sitting at the board of directors' table with her associates, caught at the moment of saying "aye" or "nay." Or he is a bespectacled gentleman, bent over a docket of papers, his pen hovering over the line marked (X).

All of these images have a significant point in common. In them, the decision maker is a person at the moment of choice, ready to plant his foot on one or another of the routes that lead from the crossroads. All the images falsify decision by focusing on its final moment. All of them ignore the whole lengthy, complex process of alerting, exploring, and analyzing that precede that final moment, and the process of evaluating that succeeds it. The images we will use in this chapter are quite different, showing the manager in all of the stages of decision, and not simply in the act of choice.

The Executive as Decision Maker

Decision making comprises four principal phases: finding occasions for making a decision, finding possible courses of action, choosing among courses of action, and evaluating past choices. These four activities account for quite different fractions of the time budgets of executives. Although the fractions vary greatly from one organization level to another and from one executive to another, we can make some generalizations about them even from casual observation. Executives and their staffs spend a large fraction of their time surveying the economic, technical, political, and social environment to identify new conditions that call for new actions. They probably spend an even larger fraction of their time, individually or with their associates, seeking to invent, design, and develop possible courses of action for handling situations where a decision is needed. They spend a small fraction of their time in choosing among alternative actions already developed to meet an identified problem and already analyzed in terms of their consequences. They spend a moderate portion of their time assessing the outcomes of past actions as part of a repeating cycle that leads again to new decisions. The four fractions, added together, account for most of what executives do.[1]

The first phase of the decision-making process—searching the environment for conditions calling for decision—I shall call

[1] The way in which these activities take shape within an organization is described in some detail in James G. March and Herbert A. Simon, *Organizations* (New York: John Wiley & Sons, Inc., 1958), chaps. 6 and 7. A wealth of information about the allocation of managerial time and attention may be found in Henry Mintzberg, *The Nature of Managerial Work* (New York: Harper & Row, 1973).

intelligence activity (borrowing the military meaning of intelligence). The second phase—inventing, developing, and analyzing possible courses of action—I shall call *design* activity. The third phase—selecting a particular course of action from those available—I shall call *choice* activity. The fourth phase, assessing past choices, I shall call *review* activity. Most of the discussion here will be directed to the first three phases.

the phases of decision: one example

Let me illustrate these phases of decision with an historical example that pertains directly to a central topic of this book: the introduction of modern computers into the management process, and the progressive development of computer systems in business and governmental organizations. How did this come about? Computers first became available commercially in the early 1950s.[2] Although, in some vague and general sense, company managements were aware that computers existed, few managements had investigated their possible applications with any thoroughness before about 1955. For most companies, the use of computers required no decision before that time because it hadn't been placed on the agenda.[3]

The intelligence activity preceding the initial introduction of computers tended to come about in one of two ways. Some companies—for example, in the aerospace industries—were burdened with enormously complex computations for engineering design. Because efficiency in computation was a constant problem, and because the design departments were staffed with engineers who could understand, at least in general, the technology of computers, awareness of computers and their potentialities came early to these companies. After computers were already used extensively for design calculations, businesses with a large data-

---

[2] A prediction I published in 1950 that computers would be used widely in management was received with utmost skepticism by the managers who heard or read it. Not all of my predictions have been realized so promptly. See my "Modern Organization Theories," *Advanced Management,* 15 (1950), 2, 4.

[3] Richard M. Cyert, Herbert A. Simon, and Donald B. Trow, "Observation of a Business Decision," *Journal of Business,* 29 (1956), 237-48.

processing load—insurance companies, accounting departments in large firms, banks—discovered, or were introduced to, these new devices and began to consider seriously their possible introduction.

Once it was recognized that computers might have a place in modern business, a major design task had to be carried out in each company before they could be introduced. It is now a commonplace that payrolls can be prepared by computers; to design a program for this purpose is a routine matter. It may not be remembered that the first such program was successfully completed only about twenty years ago, and that writing and installing it was a major undertaking that several times hovered on the brink of failure. (A General Electric plant in Cincinnati is usually credited with the first successful payroll program.) Often the crucial step in the introduction of a computer into a company was the initial step of putting the question on the agenda: Few companies that carried their investigations of computers to the point where they had definite plans for a major possible application failed to install them. Commitment to the new course of action took place gradually, but usually irreversibly, as the intelligence and design phases of the decision were going on. The final choice was, in many instances, almost *pro forma*.

As we know so well, the introduction of a first computer into a company did not end the decision process. Actual experience with the new technology convinced most managements that it was here to stay. The new computer brought with it new experts and technicians to program and operate it, and new knowledge and sophistication about its capabilities among engineers, accountants, and other users. It opened the window of the company to the burgeoning world of computer science, and to the new potentialities that were being demonstrated in application throughout the business and governmental worlds. Bringing the first computer into a company was not only a decision; it was also a major piece of intelligence activity. By increasing awareness of computers, it provided new occasions for decisions about potential applications. The process, of course, has not yet come to an end.

On some early occasions when I was asked to advise companies about the acquisition of a computer, my advice was that, before they made any decision or commitment, they should make a careful determination of whether they needed such a device and how they would use it. I soon realized that this was poor advice—that a company only acquired the ability to make sound decisions about computers (or almost any other very novel technology, for that matter) by hands-on experience with them. The first investment in a computer, preferably one of modest size, was not to be judged by its cost-saving potential—it might have none—but by its contribution to intelligence capabilities and subsequent decisions.

interweaving of the phases

Generally speaking, intelligence activity precedes design, and design activity precedes choice. The cycle of phases is, however, far more complex than this sequence suggests. Each phase in making a particular decision is itself a complex decision-making process. The design phase, for example, may call for new intelligence activities; problems at any given level generate subproblems that, in turn, have their intelligence, design, and choice phases, and so on. There are wheels within wheels within wheels.[4] Nevertheless, the three major phases are often clearly discernible as the organizational decision process unfolds. They are closely related to the stages in problem solving first described by John Dewey:

> What is the problem?
> What are the alternatives?
> Which alternative is best? [5]

In the foregoing discussion I have ignored the fourth phase of decision making: the task of carrying out decisions. I shall merely observe by the way that seeing that decisions are executed is again decision-making activity. A broad policy deci-

[4] See my "The Structure of Ill-Structured Problems," *Artificial Intelligence*, 4 (1973), 181-202.
[5] John Dewey, *How We Think* (New York: D. C. Heath & Company, 1910), chap. 8.

sion creates a new condition for the organization's executives that calls for the design and choice of a course of action for executing the policy. Executing policy, then, is indistinguishable from making more detailed policy. For this reason, I shall feel justified in taking my pattern for decision making as a paradigm for most executive activity.

developing decision-making skills

It is an obvious step from the premise that managing is decision making to the conclusion that the important skills for an executive are decision-making skills. It is generally believed that good decision makers, like good athletes, are born, not made. The belief is about the same half-truth in the one case as it is in the other. That human beings come into the world endowed unequally with biological potential for athletic prowess is undeniable. They also come endowed unequally with intelligence, cheerfulness, and many other characteristics and potentialities. To a limited extent, we can measure some aspects of that endowment—height, weight, perhaps intelligence. Whenever we make such measurements and compare them with adult performance, we obtain significant, but low, correlations. A person who is not a natural athlete is unlikely to run the four-minute mile; but many natural athletes have never come close to that goal. A person who is not naturally intelligent is unlikely to star in science; but many intelligent scientists are not stars.

A good athlete is born when someone with some natural endowment, by dint of practice, learning and experience develops that natural endowment into a mature skill. A good executive is born when someone with some natural endowment (intelligence, vigor and some capacity for interacting with his fellow men) by dint of practice, learning and experience develops that endowment into a mature skill. The skills involved in intelligence, design and choosing activities are as learnable and trainable as the skills involved in driving, recovering and putting a golf ball. I hope to indicate in the course of this chapter some of the things a modern executive needs to learn about decision making.

executive responsibility for organizational decision making

The executive's job involves not only making decisions himself, but also seeing that the organization or the part of an organization that he directs makes decisions effectively. The vast bulk of the decision-making activity for which he is responsible is not his personal activity, but the activity of his subordinates.

Nowadays, with computers everywhere, we can think of information as something almost tangible: strings of symbols which, like strips of steel or plastic ribbons, can be processed—changed from one form to another. We can think of white-collar organizations as factories for processing information. The executive is the factory manager, with all the usual responsibilities for maintaining the factory operation, getting it back into production when it breaks down, and proposing and carrying through improvements in its design.

There is no reason to expect that a person who has acquired a fairly high level of skill in decision-making activity will have a correspondingly high skill in designing efficient decision-making systems. To imagine that there is such a connection is like supposing that a man who is a good weight lifter can therefore design cranes. The skills of designing and maintaining the modern decision-making systems we call organizations are less intuitive skills. Hence, they are even more susceptible to training than the skills of personal decision making.

programed and nonprogramed decisions

In discussing how executives now make decisions, and how they will make them in the future, let us distinguish two polar types of decisions. I shall call them *programed decisions* and *nonprogramed decisions*, respectively.[6] Having christened them, I

---

[6] Often the alternative terms "well-structured" and "ill-structured" are used in place of "programed" and "nonprogramed," but since the latter terms have been in the management literature for fifteen years or more and are widely used, I will stick to them here.

hasten to add that they are not really distinct types, but a whole continuum, with highly programed decisions at one end of that continuum and highly unprogramed decisions at the other end. We can find decisions of all shades of gray along the continuum, and I use the terms programed and nonprogramed simply as labels for the black and the white of the range.[7]

Decisions are programed to the extent that they are repetitive and routine, to the extent that a definite procedure has been worked out for handling them so that they don't have to be treated *de novo* each time they occur. The obvious reason why programed decisions tend to be repetitive, and vice versa, is that if a particular problem recurs often enough, a routine procedure will usually be worked out for solving it. Numerous examples of programed decisions in organizations will occur to you: pricing ordinary customers' orders; determining salary payments to employees who have been ill; reordering office supplies.

Decisions are nonprogramed to the extent that they are novel, unstructured and unusually consequential. There is no cut-and-dried method for handling the problem because it hasn't arisen before, or because its precise nature and structure are elusive or complex, or because it is so important that it deserves a custom-tailored treatment. The decision of a company to establish operations in a country where it has not been before is a good example of a nonprogramed decision. Remember, we are considering not merely the final act of approval of the step, but the whole complex of intelligence and design activities that preceded it. Many of the component activities were no doubt programed—employing standard business techniques—but before these components could be designed and assembled they had to be provided with a broader framework of corporate strategy.

I have borrowed the term program from the computer trade, and intend it in the sense in which it is used there. A *program* is a detailed prescription or strategy that governs the sequence of responses of a system to a complex task environment. Most of the programs that govern organizational response are not as de-

---

[7] See March and Simon, *Organizations*, pp. 139-42 and 177-80 for further discussion of these types of decisions. There we contrasted "routine" decisions with decisions requiring problem solving.

tailed or as precise as computer programs. However, they all have the same intent: to permit an adaptive response of the system to the situation.

In what sense, then, can we say that the response of a system to a situation is nonprogramed? Surely something determines the response. That something, that collection of rules of procedure, is by definition a program. By nonprogramed I mean a response where the system has no *specific* procedures to deal with a situation like the one at hand, but must fall back on whatever *general* capacity it has for intelligent, adaptive, problem-oriented action. In addition to his specific skills and specific knowledge, man has some general problem-solving capacities. Given almost any kind of situation, no matter how novel or perplexing, he can begin to reason about it in terms of ends and means.

The general problem-solving equipment is not always effective. Men often fail to solve problems, or they reach unsatisfactory solutions. But man is seldom completely helpless in a new situation. He possesses general problem-solving equipment which, however inefficient, fills some of the gaps in his special problem-solving skills. And organizations, as collections of men, have some of this same general adaptive capacity.

The cost of using general-purpose programs to solve problems is usually high. It is advantageous to reserve these programs for situations that are truly novel, where no alternative programs are available. If any particular class of situations recurs often enough, a special-purpose program can be developed that gives better solutions and gives them more cheaply than the general problem-solving apparatus.

The main reason for distinguishing between programed and nonprogramed decisions is that different techniques are used for handling these two aspects of our decision making. The distinction, then, will be a convenient one for classifying these techniques. I shall use it for that purpose, hoping that the reader will remind himself from time to time that the world is mostly gray with only a few patches of pure black or white.

The four-fold table below (Figure 1) will provide a map of the territory I propose to cover. In the northern half of the map are some techniques related to programed decision making; in the southern half, some techniques related to nonprogramed de-

FIGURE 1. TRADITIONAL AND MODERN TECHNIQUES OF DECISION MAKING

Types of Decisions	Decision-Making Techniques	
	Traditional	*Modern*

Programed:
 	
Routine, repetitive decisions
 	
Organization develops specific processes for handling them
	1. Habit
	2. Clerical routine:
	    Standard operating procedures
	3. Organization structure:
	    Common expectations
	    A system of subgoals
	    Well-defined informational channels
		1. Operations Research:
		    Mathematical analysis
		    Models
		    Computer simulation
		2. Electronic data processing

Nonprogramed:
 	
One-shot, ill-structured novel, policy decisions
 	
Handled by general problem-solving processes
	1. Judgment, intuition, and creativity
	2. Rules of thumb
	3. Selection and training of executives
		Heuristic problem-solving techniques applied to:
		    (a) training human decision makers
		    (b) constructing heuristic computer programs

48

cision making. In the western half of the map I placed the classical techniques used in decision making—the kit of tools that has been used by executives and organizations from the time of the earliest recorded history up to the present generation. In the eastern half of the map I placed the new techniques of decision making—tools that have been forged largely since World War II, that have only recently come into extensive use in management in this country, and that are still experiencing vigorous development. I shall proceed across the map from north to south, and from west to east, taking up, in order, the northwest and the southwest quadrants, the northeast quadrant, and the southeast quadrant.

The western, traditional, half of the map is more familiar to us than the eastern. Moreover, the northeast quadrant is largely concerned with the tools of operations research that have now been with us in programed decision making for twenty years. It is more familiar than the southeast—which deals with heuristic problem-solving techniques for nonprogramed decision making that are still relatively novel and undeveloped. For these reasons, we will pay most attention to the eastern half, and particularly the southeast quadrant, which is the most problematic and speculative.

Traditional Decision-Making Methods

Let us examine the western half of our map of decision-making techniques (Fig. 1). This half represents methods that have been widely understood and applied in human organizations at least from the time of the building of the pyramids. In painting with a broad brush, I may convey the impression that there was no progress in organizational matters during the course of three millennia. I do not believe this to be true, and I do not intend to imply it, but the progress that was made did not enlarge the repertory of basic mechanisms to which I shall refer.

We shall consider, in turn, techniques for making programed decisions (northwest quadrant) and techniques for making nonprogramed decisions (southwest quadrant).

traditional techniques for programed decisions

"Man," says William James, "is born with a tendency to do more things than he has ready-made arrangements for in his nerve centres. Most of the performances of other animals are automatic. But in him the number of them is so enormous that most of them must be the fruit of painful study. If practice did not make perfect; nor habit economize the expense of nervous and muscular energy, he would therefore be in a sorry plight." [8]

Habit is the most general, the most pervasive, of all techniques for making programed decisions. The collective memories of organization members are vast encyclopedias of factual knowledge, habitual skills, and operating procedures. The large costs associated with bringing new members into organizations are principally costs of providing the new members, through formal training and experience, with the repertory of skills and other habits they need in their jobs. Partly, the organization provides these habits; partly, it acquires them by selecting new employees who have already learned them in the educational and training institutions that society maintains.

Closely related to habits are standard operating procedures. The only difference between habits and standard operating procedures is that the former have become internalized—recorded in the central nervous system—while the latter begin as formal, written, recorded programs. Standard operating procedures provide a means for indoctrinating new members into the habitual patterns of organizational behavior, a means for reminding old members of patterns that are used so infrequently that they never become completely habitual, and a means for bringing habitual patterns out into the open where they can be examined, modified and improved.

Organization structure, over and above standard operating procedures, is itself a partial specification of decision-making programs. The organization structure establishes a common set of presuppositions and expectations as to which members of the

[8] William James, *The Principles of Psychology* (New York: Henry Holt & Company, 1890) or (New York: Dover Publications, Inc., 1950), 1, 113.

organization are responsible for which classes of decisions; it establishes a structure of subgoals to serve as criteria of choice in various parts of the organization; and it establishes intelligence responsibilities in particular organization units for scrutinizing specific parts of the organization's environment and for communicating events requiring attention to appropriate decision points. In the past, the improvement of programed decision making in organizations has focused largely upon these techniques: upon improving the knowledge, skills and habits of individual employees by means of training programs and planned tours of duty; upon developing better standard operating procedures and securing adherence to them; and upon modifying the structure of the organization itself, the division of labor, the subgoal structure, the allocation of responsibilities.

Mankind has possessed for many centuries an impressive collection of techniques for developing and maintaining predictable programed responses in an organization to those problems posed by its environment that are relatively repetitive and well structured. The history of the development of these techniques has never been written—much of it is undoubtedly buried in prehistory—but one can point to particular periods of innovation. The scientific management movement beginning at the turn of the century, and particularly the development of standard methods for performing repetitive work, is one of the most recent of these.

traditional techniques for nonprogramed decisions

When we turn to the area of nonprogramed decisions, we have much less to point to in the way of specific, describable techniques. When we ask how executives in organizations make nonprogramed decisions, we are generally told that they "exercise judgment," and that this judgment depends, in some undefined way, upon experience, insight and intuition. If the decision we are inquiring about was a particularly difficult one, or one that yielded especially impressive results, we may be told that creativity was required.

There is a scene in Moliere's *Le Malade Imaginaire* in which the physician is asked why opium puts people to sleep. "Because it possesses the dormitive faculty," he replies triumphantly. To name a phenomenon is not to explain it. Saying that nonprogramed decisions are made by exercising judgment *names* that phenomenon but does not explain it. It doesn't help the man who lacks judgment (i.e., who doesn't make decisions well) to acquire it.

Making programed decisions depends on relatively simple psychological processes that are somewhat understood, at least at the practical level. These include habit, memory, simple manipulations of things and symbols. Making nonprogramed decisions depends on psychological processes that, until recently, have not been understood at all. Because we have not understood them, our theories about nonprogramed decision making have been rather empty and our practical advice only moderately helpful.

One thing we have known about nonprogramed decision making is that it can be improved somewhat by training in orderly thinking. In addition to the very specific habits one can acquire for doing very specific things, one can acquire the habit —when confronted with a vague and difficult situation—of asking, "What is the problem?" We can even construct rather generalized operating procedures for decision making. The military "estimate of the situation"—a checklist of things to consider in analyzing a military decision problem—is an example of such an operating procedure.

There is nothing wrong with such aids to decision making except that they don't go nearly far enough. They graduate the decision maker from nursery school to kindergarten, perhaps, but they don't carry his education much further.

How then do executives discharge their responsibilities for seeing that decision making in their organizations, nonprogramed as well as programed, is of high quality? Let me propose an analogy. If you have a job to do, and you don't have the time or the skill to design and produce just the right tool to do it, you look around among the tools you have or can get at the hardware store and select the best one you can find. We haven't known very much about how to improve human decision-making skills, but we observe that some people have these skills much better developed than others. We try to hire those people. We rely on personnel selection as our principal technique for improving complex decision-making skills in organizations.

We supplement our selection techniques with two kinds of training: the professional training in basic principles that generally precedes entrance into organizational life, and the training through experience and planned job rotation that the organization itself can provide. Sometimes we supplement the latter with advanced management training in a university setting or a company training program.

A person is usually able to improve his decision-making skills gradually through appropriate experience, even if he himself has only the sketchiest notion of how he exercises those skills. Man is a learning animal. If he is subjected to a sequence of problem situations of progressively greater difficulty and of difficulty appropriate to the level of skill he has attained, he will usually show an increasing capacity to handle problems of those kinds well. For problems of a nonprogramed sort neither he nor we have known from whence the improvement comes. The processes of learning have been as mysterious as the processes of problem solving. But improvement there is. We have thus been able, in a crude way, to use training and planned experience as a means for improving nonprogramed decision making in organizations.

Appropriate design of the organization structure may facilitate nonprogramed as well as programed decision making. An important principle of organization design has been called facetiously "Gresham's Law of Planning." It states that programed activity tends to drive out nonprogramed activity. If an executive has a job that involves a mixture of programed and nonprogramed decision-making responsibilities, the former will come to crowd out the latter. The organizational implication of Gresham's Law is that special provision must be made for nonprogramed decision making by creating specific organizational responsibilities and organizational units to take care of it. The various kinds of staff units that are so characteristic of large-scale modern organizations are mostly units specialized in particular aspects of the more complex nonprogramed decision-making tasks. Market research units and research departments, to cite some examples, specialize in the intelligence phase of decision making; planning departments and product development departments specialize in the design phase. The creation of organizational units to carry on these activities allocates brainpower to nonprogramed thought and provides some minimal assurance that such thought will occur in the organization.

In summary, we have not had, in the past, adequate knowledge of the processes that are involved in decision making in complex situations. Human thinking, problem solving and learning have been mysterious processes which we have labeled but not explained. Lacking an understanding of these processes, we have had to resort to gross techniques for improving nonprogramed decision making: selection of men who have demonstrated their capacity for it; further development of their powers through professional training and planned experience; protection of nonprogramed activity from the pressure of repetitive activity by establishing specialized organizational units to carry it on. We cannot say that these traditional techniques failed—decisions are made daily in organizations. Neither can we say that we might not do very much better in the future as our knowledge of the decision-making process grows.

New Techniques for Programed Decision Making

World War II brought large numbers of scientists trained in the use of mathematical tools into contact, for the first time, with operational and managerial problems. Designers of military aircraft could not plan aircraft armament without making assumptions about the formations in which the planes would be flown and the strategy of their commitment to action. Mathematical economists responsible for material allocation had to come to grips with complex logistics systems. The need for solving these problems, coupled with the tools of quantitative analysis that the scientists and econometricians brought with them, have produced some new approaches to management decision that have brought about fundamental changes in the methods used for programed decision making.[9]

operations research

The terms "operations research" and "management science" are nowadays used almost interchangeably to refer to the application of orderly analytic methods, often involving sophisticated mathematical tools, to management decision making, and particularly to programed decision making. At a more philosophic level, operations research may be viewed as the application of scientific method to management problems, and in this sense simply to represent a continuation of the earlier scientific management movement. Charles Babbage and Frederick Taylor will have to be made, retroactively, charter members of the operations research societies.

Historically, operations research and management science did not in fact emerge out of scientific management or industrial engineering. As a sociological movement, operations research, emerging out of the military needs of World War II, brought the decision-making problems of management within the range of interests of large numbers of natural scientists, and particularly of mathematicians and statisticians.[10] The operations researchers soon joined forces with mathematical economists who had come into the same area—to the mutual benefit of both groups. And there was soon widespread fraternization between these exponents of the "new" scientific management and men trained in the earlier traditions of scientific management and industrial engineering. No meaningful line can be drawn any more to demarcate operations research or management science from the two older fields.

[9] See Fig. 1, p. 48, northeast quadrant.
[10] Two standard early works in operations research by leading pioneers in the movement are C. West Churchman, Russell L. Ackoff, and E. Leonard Arnoff, *Introduction to Operations Research* (New York: John Wiley & Sons, Inc., 1957); and Philip M. Morse and George E. Kimball, *Methods of Operations Research* (New York: John Wiley & Sons, Inc., 1951). The principal professional journals are *Operations Research,* published by the Operations Research Society of American, and *Management Science,* published by The Institute of Management Science.

Along with some mathematical tools, which I shall discuss presently, operations research brought into management decision making a point of view called the systems approach. The systems approach is a set of attitudes and a frame of mind rather than a definite and explicit theory. At its vaguest, it means looking at the whole problem—again, hardly a novel idea and not always a very helpful one. Somewhat more concretely, it means designing the components of a system and making individual decisions within it in the light of the implications of these decisions for the system as a whole.[11] We now know a *little* about how this might be done:

1. Economic analysis has something to say about rational behavior in complex systems of interacting elements, and particularly about the conditions under which the choices that are optimal for subsystems will or will not be optimal for a system as a whole. Economic analysis also has a great deal to say about the price system as a possible mechanism for decentralizing decision making without ignoring the interactions among subsystems.[12]

2. Mathematical techniques have been developed and adapted by engineers and economists for analyzing the dynamic behavior of complex systems. Under the labels of servomechanism theory and cybernetics, such techniques underwent rapid development at about the time of World War II. They are frequently referred to now under the more general name of "dynamic programing." Such techniques have considerable usefulness in the design of dynamic systems.[13]

[11]See Churchman, *et al.*, *Introduction*, pp. 109-11.

[12] Much contemporary economic analysis now employs the operations research tool of linear programing for the description and analysis of market systems. The pioneering work here was Tjalling C. Koopmans, ed., *Activity Analysis of Production and Allocation* (New York: John Wiley & Sons, Inc., 1951).

[13] The word "cybernetics" was first used by Norbert Wiener in *Cybernetics* (New York: John Wiley & Sons, Inc., 1948), p. 19. A good example of the application of dynamic programing techniques to an industrial programed decision problem is Charles C. Holt, Franco Modigliani, John F. Muth, and Herbert A. Simon, *Planning Production, Inventories, and Work Force* (Englewood Cliffs, N.J.: Prentice-Hall, Inc., 1960). Since "systems design" is a very modish, if not faddish, word, I don't want to exaggerate the amount

the mathematical tools

Operations research progressed from the talking to the action stage by finding tools with which to solve concrete managerial problems. Among the tools, some of them relatively new, some of them already known to statisticians, mathematicians or economists were linear programing, dynamic programing, integer programing, game theory, Bayesian decision theory, queuing theory, and probability theory. Behind each of these formidable terms lies a mathematical model for a range of management problems. Linear programing, for example, can be used to provide a mathematical model for the operations of a gasoline refinery, or a commercial cattle-feed manufacturing operation. Dynamic programing can be used as a model for many inventory and production planning situations. Integer programing is applicable to planning and scheduling problems that must have discrete rather than continuous solutions. Game theory models have been used to represent marketing problems. Queuing theory has been widely used to handle scheduling tasks and other problems involving waiting lines. Bayesian decision theory provides models for making choices among alternatives under uncertainty about the outcomes. Probability theory is a component of several of these tools, and has been used directly in a wide variety of contexts—it has been, perhaps, the most versatile of all the tools.

Whatever the specific mathematical tool, the general recipe for using it in management decision making is something like this:

1.   Construct a *mathematical model* that satisfies the conditions of the tool to be used and which, at the same time, mirrors the important factors in the management situation to be analyzed. For success, the basic structure of the tool must fit the basic structure of the problem, although compromise and

---

of well-understood technique that stands behind it. Nevertheless, it is fair to say that we can approach the design and analysis of large dynamic systems today with a good deal more sophistication than we could ten or twenty years ago.

approximation is often necessary in order to fit them to each other.

2. Define a *criterion function*, a measure to be used for comparing the relative merits of various possible courses of action.

3. Obtain *empirical estimates* of the numerical parameters in the model that specify the particular, concrete situation to which it is to be applied.

4. Carry through the *mathematical calculations* required to find the course of action which, for the specified parameter values, maximizes the criterion function. With each of the tools are associated computational procedures for carrying out these calculations more or less efficiently.

In any decision-making situation where we apply this recipe successfully, we have, in fact, constructed a *program* for the organization's decisions. We have either annexed some decisions that had been judgmental to the area of programed decision making,[14] or we have replaced a rule-of-thumb program with a more sophisticated program that guarantees us optimal decisions —optimal, that is, within the framework of the mathematical model.

But certain conditions must be satisfied in order to apply this recipe to a class of decision problems. First, mathematical variables must be defined to represent the important aspects of the situation. In particular, a quantitative criterion function must be defined. If the problem area is so hopelessly qualitative that it cannot be described even approximately in terms of such variables, the approach fails. Second, parameters of the model's structure have to be estimated before the model can be applied in a particular situation. Hence, there must exist ways of making actual numerical estimates of these parameters—estimates of sufficient accuracy for the practical task at hand. Third, the specification of the model must fit the mathematical tools to be used. If certain kinds of nonlinearities are absolutely crucial to an

---

[14] Thus, operations research in addition to providing techniques for programed decisions also expands the boundaries of programed decision making into areas that were previously nonprogramed.

accurate description of the situation, linear programing simply won't work—it is a tool that is limited to mathematical systems that are, in a certain sense, linear. Fourth, the problem must be small enough that the calculations can be carried out in reasonable time and at a reasonable cost, although, of course, the computer has enabled us to handle immensely larger problems than we were able to handle without it.

Some relatively straightforward management problems—for example, many problems of factory scheduling—turn out to be far too large for even such a powerful tool as integer programing, and even when that tool is implemented with computers. It is easy for the operations research enthusiast to underestimate the stringency of the conditions for applicability of his methods. This leads to an ailment that might be called mathematician's aphasia. The victim abstracts the original problem until the mathematical or computational intractibilities have been removed (and all semblance of reality lost), solves the new simplified problem, and then pretends that this was the problem he wanted to solve all along. He hopes the manager will be so dazzled by the beauty of the mathematical formulation that he will not remember that his practical operating problem has not been handled.

It is just as easy for the traditionalist to overestimate the stringency of the conditions. For the operations research approach to work, nothing needs to be exact—it just has to be close enough to give better results than could be obtained by common sense without the mathematics, and that is often not a difficult criterion to beat. Furthermore, it is dangerous to assume that something is essentially qualitative and not reducible to mathematical form until an applied mathematician has had a try at it. For example, when designing dynamic programing schemes for handling inventory planning, I have often been told by my company clients that "you can't place a dollar value on a lost order from inventory runout." Of course you can't, but why, the answer goes, can't you estimate the penalty cost of taking emergency action to *avoid* losing the order—shipping, for example, by air express? Thus, many things that seem intangible and inherently qualitative can be reduced, for management decision-making purposes, to dollars and cents.

But we need not spin out these generalities. Operations

research techniques are now applied in a vast number of practical management situations. In many of these situations, when mathematical techniques were first proposed there was much head shaking and muttering about judgment. The area of application is large. It continues to grow. But there is no indication that it will cover the whole of management decision making. Difficulties in quantifying set one boundary; limits on computing power set another. Although the boundaries are movable, they have a long way to go before they will encompass all of management.

enter the computer

It was a historical accident with large consequences that the same war that spawned operations research saw also the birth of the modern digital computer as a practical device for management.[15] The computer was conceived as a means for exploring by numerical analysis the properties of mathematical systems too large or too complex to be treated by known analytic methods. The systems of differential equations that were arising in aerodynamics, meteorology, and the design of nuclear reactors were obvious candidates for this treatment. It was soon realized that equally large or larger problems were generated by the linear programing and dynamic programing models of management decision problems. Whatever the conceptual power of the mathematical models of operations research, their actual use in practical schemes for decision making hinged on the fortuitous arrival on the scene of the computer.

While computers were initially conceived for doing arithmetic on problems that had first been cast in a mathematical form having known solution procedures, it gradually became clear that there were other ways of using them. If a model or simulation of a situation could be programed for a computer, the behavior of the system could then be studied simply by having the computer imitate it—and without solving, in the traditional

---

[15] A general book on the history of the development of computers and on their use by management is John A. Postley, *Computers and People* (New York: McGraw-Hill Book Company, Inc., 1960).

analytic sense, the mathematical equations themselves. Of course, there is a little more to simulation than this. In general, we need to simulate the behavior of the system not under a single set of conditions but over a whole range of conditions. Having simulated it, we need some procedure for evaluating the results—for deciding whether the system behavior was satisfactory or not. Finally, before we can simulate the behavior at all, we have to estimate accurately enough the structure of the system—simulation techniques do not reduce the burden of providing numerical estimates of system parameters.

In spite of these limitations and difficulties, simulation has enabled airlines to determine how many reserve aircraft they should keep on hand, has been used to study highway congestion, has led to improvement in inventory control procedures for large warehousing operations, and has accomplished many other difficult tasks. For usefulness, simulation does not need to find optimal solutions; it only has to do better than common sense and judgment, and in many situations it can do just that.

Of course, the bread-and-butter applications of computers to business decision making have had little to do with either mathematical models or simulation. They have had to do with automating a whole host of routine and repetitive data-processing activities in factory and office that had for many years been highly programed but not nearly so completely automated. Through this development, large-scale data processing has itself become a factory operation, automated to a degree exceeding all but a very few manufacturing processes.

the revolution in programed decision making

The revolution in programed decision making has by no means reached its limits, but we can see its shape. The rapidity of change stems partly from the fact that there has been not a single innovation but several related innovations, all of which contribute to it.

1.   The electronic computer has brought about, swiftly, a

high level of automation in the routine, programed decision making and data processing that were formerly the province of clerks.

2. The area of programed decision making is being rapidly extended as we find ways to apply the tools of operations research to more and more types of decisions previously regarded as judgmental—particularly, but not exclusively, middle-management decisions in the areas of manufacturing and warehousing.

3. The computer has extended the applicability of the mathematical techniques to problems far too large to be handled by less automatic computing devices, and has further extended the range of programable decisions by contributing the new technique of simulation.

4. In the past ten years, companies have begun to bring together the first two of these developments: combining the mathematical techniques for making decisions about aggregative middle-management variables with the data-processing techniques for implementing these decisions in detail at clerical levels. Increasingly, the same nearly automatic system that fills orders, sends invoices, and records payments also takes care of calculating rules for optimal inventory levels and reorder quantities.

Out of the combination of these four developments there has emerged the new picture of a data-processing factory for manufacturing, automatically, the organization's programed decisions—just as the physical processing factory manufactures its products in a manner that becomes increasingly mechanized. The nearly automated factory operates on the basis of programed decisions produced in the nearly automated office beside it.

Heuristic Problem Solving

However significant the techniques for programed decision making that have emerged over the past two decades, and however great the progress in reducing to sophisticated programs some areas that had previously been unprogramed, these developments still leave untouched a major part of managerial decision-making activity. Many, perhaps most, of the problems that have to be handled at middle and high levels in management have not been made amenable to mathematical treatment, and probably never will.

But that is not the whole story. The processes of non-programed decision making are beginning to undergo as fundamental a revolution as the one that is currently transforming programed decision making in business organizations. Basic discoveries have been made about the nature of human problem solving, and their first potentialities for business application have already emerged. We may expect these discoveries to have increasing effects on the ways in which nonprogramed decisions are made in business organizations.[16]

There are several conceivable ways in which the limitation of the new approaches to programed decision making might be overcome. One of these would be to discover how to increase substantially the problem-solving capabilities of humans in non-programed situations. Another way would be to discover how to use computers to aid humans in problem solving without first reducing the problems to mathematical or numerical form.

Both of these possibilities hinge on deepening our understanding of human problem-solving processes. If we understand how something is accomplished, and the processes that are involved, we can either try to improve those processes or find alternatives to them. Of course, we might invent synthetic processes without understanding how the natural process works. Airplanes were invented before we understood the flight of birds. And linear programing was applied to refinery scheduling before we knew what mental processes humans had used to schedule refineries. Nevertheless, when we run out of ideas for handling poorly structured problem-solving tasks, it seems plausible to examine more closely the processes used by humans who have handled such tasks—not always efficiently, to be sure—for several millennia.

This section is concerned with what is now being learned, through research, about human problem solving and about artificial intelligence techniques for handling nonprogramed decision

---

[16] See Figure 1, p. 48, southeast quadrant.

making. Since these developments still remain largely, but not exclusively, in the basic research laboratory, only a few concrete examples can be provided at present to show the specific shape of their management applications. The main purpose of these pages, then, is to describe this important set of new discoveries about information processing in man and computer that are likely to have, over the next generation, at least as large an impact on management as the older operations research ideas have had during the generation just past. The final section of the chapter will speculate on the nature of this largely prospective impact.

understanding human
problem-solving processes

It is only in the past twenty years that we have begun to have a good scientific understanding of the information processes that humans use in problem solving and nonprogramed decision making. Of course some things have been known about problem solving for a long time, from common sense observation of what goes on around us (and in our own heads, too). We have known, for example, that problem solving usually involves a great deal of search activity of one sort or another, that it often uses abstraction and imagery, that small hints can have dramatic effects on the ease of solution of a problem, and so on. But all the processes that were observed in casual ways—particularly the search activities and the use of relatively obvious perceptual clues—appeared so simple that we did not believe they could account for the impressive outcomes. The achievements of the problem-solving process—the bridges it designs, the organizations it builds and maintains, the laws of nature it discovers, have an impressiveness all out of proportion to the groping, almost random, processes that we observe in the problem solver at work. Little wonder that we invented terms like *intuition, insight,* and *judgment,* and invested them with the mystery of the whole process.

In recent years we have made our observations more systematically. We give a subject a problem—say, proving a theorem in Euclidean geometry—and ask him to think aloud while he solves it. We have no illusion that all his thought processes will rise to the level of consciousness or be verbalized, but we hope to get some clues about the course his thought takes. We tape record what he says during the ten or fifteen minutes he works on the problem.[17]

From the tape recording, we observe that the subject compares the theorem to be proved with some theorems he knows—he looks for similarities and differences. These suggest subproblems whose solution may contribute to the solution of the main problem: "I have to prove two triangles congruent. Are any pairs of sides equal? Can I prove some pairs of sides equal?" Subproblems may, in turn, generate new subproblems until he comes to a problem he can solve directly. Then he climbs back up to the next level of problems above. He gradually begins to assemble results that look as though they will contribute to the solution of the whole problem. He persists along a path as he gets warmer, backs off to another direction of search when he finds a particular trail getting cooler.

At one level, nothing seems complicated here—nothing is very different from the white rat in the laboratory sniffing his way through a maze. But still the feeling persists that we are seeing only the superficial parts of the process—that there is a vast iceberg underneath, concealed from our view and from the consciousness of the subject.

We have a good deal of evidence today that this feeling of mystery is an illusion. It would seem that the subconscious parts of the process are no different in kind from the parts we observe. We learn that the complexity of the problem-solving process that makes its outcomes so impressive is a complexity assembled out of relatively simple interactions among a large number of extremely simple basic elements.

Even if the plausibility of such a hypothesis is admitted, how is it to be tested? The primary test has been to construct a synthetic thought process in the form of a computer program, present a problem to the program, and compare its protocol

---

[17] In *Human Problem Solving* (Englewood Cliffs, N.J.: Prentice-Hall, 1972) Allen Newell and I present and analyze a substantial body of data on problem solving, obtained by thinking-aloud methods, and we set forth the theory of problem solving that is described briefly in these pages. Another and less technical account of the theory may be found in my *The Sciences of the Artificial* (Cambridge, Mass.: M.I.T. Press, 1969).

(i.e., the computer trace) with the thinking-aloud protocol of a human problem solver given the same problem. If the synthetic thought process matches, step by step, each element in the verbalized part of the human thought process—notices the same clues in the situation, generates the same subproblems, draws from memory the same theorems—then we are justified in concluding that the computer program provides us with a basic understanding of the human process.

For twenty years now we have known how to synthesize thinking processes that parallel closely the thinking processes of human subjects in a substantial number of different problem-solving, memorizing, pattern-learning, and language-using tasks. The range of tasks that have been studied in this way is still narrow. However, little doubt remains that, in this range at least, we know what some of the principal processes of human thinking are, and how these processes are organized in problem-solving programs.

the simulation of human thought

Since the work I am describing makes important use of digital computers and of some of the central concepts associated with the invention of computers, I shall have to insert a few remarks about these devices, mentioning some of their features that are still not widely understood.

1.   There is nothing about a computer that limits its symbol-manipulating capacities to numerical symbols; computers are quite as capable of manipulating words as numbers. More literally, inside the computer there are neither words nor numbers, but simply electromagnetic patterns of some kind that, in the context of the appropriate processes for manipulating them, can be interpreted as words or numbers—or pictures or diagrams, for that matter.

2.   In principle, the potentialities of a computer for flexible and adaptive cognitive response to a task environment are no narrower and no wider than the potentialities of a human. By "in principle" I mean that the computer hardware contains these po-

tentialities, although at present we know only imperfectly how to evoke them, and we do not yet know if they are equivalent to the human capacities in speed or memory size.

Nonnumerical symbol manipulation

Historically computers were devised to do arithmetic rapidly. Certain of their components are specially adapted to this purpose. This does not mean, however, that computers can process only numerical information. All general purpose computers have the capacities to manipulate symbols, numerical or literal, in all the ways in which symbols have to be manipulated in order for them to stand for either numbers or words. Computers can read symbols from external media, copy symbols from one internal location to another, print symbols externally, compare pairs of symbols for identity or difference, associate one symbol with another, find a symbol associated with another in memory, and erase symbols. What the simulations of human thinking have shown is that complex information processing, including problem solving and decision making, can be done with just such symbol manipulations as these. Nothing more is needed.

Flexible and adaptive response

"But after all," the skeptic may say, "how can a computer be insightful or creative? It can only do what you program it to do." This statement—that computers can do only what they are programed to do—is intuitively obvious, indubitably true, and supports none of the implications that are commonly drawn from it.

A human being can think, learn, and create because the program his biological endowment gives him, together with the changes in that program produced by interaction with his environment after birth, enables him to think, learn, and create. If a computer thinks, learns, creates, it will be by virtue of a program that endows it with these capacities. Clearly this will not be a program that calls for highly stereotyped and repetitive behavior independent of the stimuli coming from the environment and the task to be completed—any more than the human's program does. It will be a program that makes the system's behavior highly conditional on the task environment—on the task goals and on the clues extracted from the environment that indicate whether progress is being made toward these goals. It will be a program that analyzes, by some means, its own performance, diagnoses its failures, and makes changes that enhance its future effectiveness. It is a simple question of fact whether a computer program can be written that will have these properties. And the answer to this question of fact is that such programs have been written.

I can now, in summary, state the central hypothesis of the theory of problem solving that has emerged from the research of the past twenty years: In solving problems, human thinking is governed by programs that organize myriads of simple information processes—or symbol manipulating processes if you like—into orderly, complex sequences that are responsive to and adaptive to the task environment and the clues that are extracted from that environment as the sequences unfold. Since programs of the same kind can be written for computers, these programs can be used to describe and simulate human thinking. Notice that the theory does not assert that there is any resemblance between the neurology of the human and the hardware of the computer. They are grossly different. However, at the level of detail represented by elementary information processes, programs can be written to describe human symbol manipulation, and these programs can be used to induce a computer to simulate the human process.

I apologize for this long methodological digression, but it is essential for an understanding of the research I shall describe that we all start from the same underlying assumptions about the nature of computers and about the nature of the phenomena we are trying to explain.

a general problem-solving program

By now, computer programs have been written to handle so many problem-solving tasks that it is impossible to enumerate all of them. There are successful programs that discover proofs for mathematical theorems, that design a number of standard industrial products (electric motors, for example), that play chess at respectable human levels, that understand and can execute

English-language instructions, that compose music, that guide a robot in the performance of simple tasks, that recognize human speech, that identify objects in visual displays or patterns in sequences of symbols, that write computer programs to perform specified tasks, that design data bases, that lay out room arrangements for houses, that diagnose illnesses. Most of these programs are the product of basic research and many have not reached the stage of application, but there are important exceptions—for example, the product design programs. The field of research that tries to explore and extend the capabilities of computers by constructing such programs is usually called artificial intelligence.

Some of these sophisticated computer programs are aimed at findings ways, whether resembling the human way or not, of doing the tasks well. Others are aimed at detailed simulation of human processes—hence, at understanding problem solving as people do it. From almost all of these efforts, whether intended as humanoid simulations or not, we learn something about human problem solving, thinking, and learning.

The first thing we have learned—and the evidence for this is by now substantial—is that these human processes can be explained *without* postulating mechanisms at subconscious levels that are different from those that are partly conscious and partly verbalized. Much of the iceberg is, indeed, below the surface and inaccessible to verbalization, but its concealed bulk is made of the same kind of ice as the part we can see. The processes of problem solving are the familiar processes of noticing, searching, and modifying the search direction on the basis of clues. The same elementary symbol-manipulating processes that participate in these functions are also sufficient for such problem-solving techniques as abstracting and using imagery. The secret of problem solving is that there is no secret. It is accomplished through complex structures of familiar simple elements. The proof is that we have been able to simulate it, using no more than those simple elements as the building blocks of our programs.

Means-ends analysis

A number of the important processes that have been observed to account for human problem solving are captured in a simulation program called GPS (General Problem Solver).[18] It is called GPS not because it can solve any kind of problem—it cannot—but because the program itself makes no specific reference to the subject matter of the problem. GPS is a program that can reason in terms of means and ends about any problem that is stated in a certain general form. Let me sketch out the basic idea:

Suppose that we are camping in the woods and decide that we need a table. How do we solve the problem of providing ourselves with one? We state the problem: We *need* a flat horizontal wooden surface; we *have* all sorts of trees around us and some tools. We ask: What is the *difference* between what we need and what we have? Trees are large, vertical cylinders of wood attached to the ground; a table top is a smaller, horizontal, movable slab of wood. Hence, there are differences in size, flatness, attachment, and so on between what we have and what we need. We ask: What tools do we have to *reduce* these differences—for example, to detach the tree from its roots? We have axes. So we apply an ax to a tree and we have solved the first subproblem—to change an object rooted in the soil into a detached object. We can now proceed in the same way to solve the second problem, making the object the right size, and the third, flattening and smoothing it.

I have, of course, vastly oversimplified the matter, but the main outlines are clear. Problem solving proceeds by erecting goals, detecting differences between present situation and goal, finding in memory or by search some tools or processes that are relevant to reducing differences of these particular kinds, and applying these tools or processes. Each problem generates subproblems until we find a subproblem we can solve—for which we already have a program stored in memory. We proceed until, by successive solution of such subproblems, we eventually achieve our overall goal—or give up. Problem solving may be viewed as a way of reaching nonprogramed decisions by reducing them to a series of programed decisions.

[18] For a fuller account of GPS, and a discussion of its validity as an explanation of human problem-solving behavior, see Newell & Simon (1972), especially chap. viii.

Means-ends analysis in GPS

The General Problem Solver mirrors this process as follows. Its programs enable it to formulate and attack three kinds of goals:

1. *Transform* goals: Change *a* into *b*.
2. *Reduce difference* goals: Eliminate or reduce the difference between *a* and *b*.
3. *Apply operator* goals: Apply the program (or operator or method) *O* to the situation *a*.

With each of these types of goals is associated one or more methods for accomplishing it. When the goal is formulated by GPS, these methods are evoked from memory and tried. A method, for example, for changing *a* into *b* is to find a difference, *d*, between them and formulate the Reduce Difference goal of eliminating this difference. A method of reducing a difference between *a* and *b* is to find an operator that is relevant for removing differences of the kind in question, and to apply that operator. A method for applying an operator is to compare the actual situation with the situation that would make it possible to apply the operator, and to formulate the goal of changing the actual situation into the required situation.

These three goal types and the methods associated with them were not manufactured out of whole cloth. They were discovered by detailed, painstaking analysis of tape recordings of the thinking-aloud protocols of subjects who were solving problems in the laboratory. With an important exception to be mentioned in a moment, almost all the goals subjects mention in the course of their efforts at solving relatively well-structured problems, and almost all the methods they apply, appear to fit the categories I have just described—although not necessarily the precise organization that has been embodied in the specific GPS structure that has been programed and run on the computer. All sorts of differences have been observed in protocols from one individual to another and from one task to another, but there is great similarity in the basic elements.

Planning in GPS

In describing GPS, I have omitted one important additional method for achieving Transform goals that was used by our subjects, and that was particularly prominent, in one form or another, in the thinking processes of the more skillful subjects. This is a method, which we call *planning*, for transforming one object into another. It works as follows: If the goal is to transform a into b, abstract a and b, eliminating most of their detail and creating the new objects ("abstractions," "images," "models") a' and b'. Now formulate the goal of transforming a' into b'. If an appropriate kind of abstraction has been used—if irrelevant detail has been discarded and the essential aspects of the situation retained, the new problem will generally be far easier to solve than the original one. Once it has been solved, it provides a series of trail markers to guide the solution of the original unabstracted problem. We were able to determine from the protocols the process of abstraction our subjects used, and to incorporate the planning method as one of the methods available to GPS.

Summary

GPS is a program—initially inferred from the protocols of human subjects solving problems in the laboratory, and subsequently coded for computer simulation—for reasoning in terms of ends and means, in terms of goals and subgoals, about problematic situations. It is subject-matter free in the sense that it is applicable to any problem that can be cast into an appropriate general form (e.g., as a problem of transforming one object into another by the application of operators). It appears to reproduce most of the processes that are observable in the behavior of the laboratory subjects in several tasks, and to explain the general organization of those processes. On the basis of simulation, we can say that GPS is a substantially correct theory of the problem-solving process as it occurs under particular laboratory conditions. Its applicability to other situations is the subject of continuing research, on which I will comment presently.

GPS does succeed in capturing some aspects of problem solving that have always been thought to be part of its mystery. For example, we can show by comparison with the human protocols that sudden insight of the "Eureka!" type sometimes occurs at the moment when the subject successfully applies the planning method and obtains a plan to guide his detailed solution. At such moments, subjects make such statements as, "Aha! I think I've got it. Just let me work it out now." The planning method also gives us a basis for exploring the processes of abstraction, which turn out, in this case at least, to involve simple and understandable information processes.

heuristic programing

Apart from their use in psychological theory, some of the ideas embedded in GPS have had practical application in artificial intelligence. Many problems are encountered in operations research work that are simply too large to be handled by the mathematical techniques (e.g., linear programing or integer programing) that are available for finding optimal solutions. Large combinatorial problems, like the problem of scheduling a job shop, often are found to be too big for the methods of mathematical optimization. In these cases, the method known as *heuristic search*, based on ideas like those incorporated in GPS, provides a powerful means for finding good, if not optimal, solutions.

Heuristic search is one example of the approach called *heuristic programing* that is now widely used in operations research and management science. There is no precise definition for the term heuristic programing, for it represents a point of view in the design of programs for complex information processing tasks, and not a precise method. This point of view is that the programs should not be limited to numerical processes, or even to orderly systematic non-numerical algorithms of the kinds familiar from the more traditional uses of computers, but that ideas should be borrowed also from the less systematic, more selective, processes that humans use in handling those many problems that have not been reduced to algorithms. It is a necessary point of view when the goal of the program writing is to simulate human thinking. It often turns out to be a useful point of view when the goal of the program writing is to supplement natural intelligence with artificial intelligence in management decision making—to bring in the computer as a problem-solving aide to the manager.

other advances in cognitive simulation

While a program like GPS answers certain important questions about human thinking and decision making, it leaves some equally important questions unanswered. At the present time, three of these unanswered questions are receiving a great deal of research attention, and substantial progress has already been made toward answering them.

The first of these research tasks is to extend the problem-solving theory from the relatively well-structured problem situations in which it was discovered and tested to problem domains that are much more poorly structured, where the problem solver's initial task is to define and represent the problem so that he can set his general problem-solving capabilities to work on it. The second of the new research questions is to explain how people understand language, and how they are able to accept problem-solving tasks stated in natural language, with all of its irregularities and rich meanings. The third research task now receiving attention is to extend the theory from puzzlelike domains, where all the information required to solve the problem is contained in the problem statement and instructions, to real-life domains where solving a problem requires the solver to draw extensively on information about the problem domain that he has stored in his long-term memory.

Ill-structured problems

When a person is given a problem to solve, unless it is of a completely familiar kind he must first determine what the problem is, understand it fully, and find some way of representing it to himself before he can go to work on it and seek a solution by heuristic search. If we think of his capability for searching out solutions as something like GPS, then he must provide that capability with a representation of the initial problem situation, of the goal situation, and of the operations available for changing the situation. He must somehow construct these representations and operations from the information in problem instructions.[19]

Using the same techniques—protocol analysis and computer simulation—that were used to develop and test GPS, J. R. Hayes and I have developed a program, which we call *Understand*, that has a modest ability to accept instructions for simple puzzles in English, to extract from them the information that would be required for a GPS-like problem solver to go to work on the problem, and to transform this information into a suitable representation of problem situation and operators.[20] The Understand program carries out its work in two discrete stages:

1. It reads and rereads the task instructions, analyzing them grammatically and semantically to extract from them their "deep structure," as linguists use that term.

2. It matches the resulting structures against a set of requirements, stored in memory, which specify in what form the structures must be placed in order to provide an acceptable input to GPS. We can think of these requirements as a set of templates that serve as a model for a well-formed problem. It then undertakes to modify the structures until they meet the requirements.

This is not the place to describe in detail how these steps are accomplished. Again, the protocol evidence we have examined shows that the Understand program follows closely the path followed by human subjects as they attempt to grasp and interpret these same task instructions. While this program, and others like it, are still in a fairly early stage of development, they begin to show us what is involved in structuring an initially unstruc-

---

[19] Recent research on providing initial structure for ill-structured problems is reported in H. A. Simon and L. Siklóssy, eds., *Representation and Meaning* (Englewood Cliffs, N.J.: Prentice-Hall, 1972), and in Lee W. Gregg, ed., *Knowledge and Cognition* (Potomac, Md.: Lawrence Erlbaum Associates, 1974).

[20] J. R. Hayes & H. A. Simon, "Understanding Written Problem Instructions," in L. W. Gregg (ed.), *Knowledge and Cognition* (Potomac, Md.: Lawrence Erlbaum Associates, 1974), pp. 167-200.

tured problem. The information processes they reveal in the human approach to these tasks show a close resemblance to the processes with which we are already familiar from earlier simulation programs.

Understanding language

The Understand program is only one of several dozen programs that perform some task requiring a degree of understanding of natural language.[21] Early approaches to computer processing of natural language were largely directed to the task of automatic language translation, and relied mostly on grammatical knowledge to analyze the sentences with which they were presented. This approach was soon found to be completely inadequate; it does not seem possible to interpret natural language without making use of considerable information about its semantic content. The more recent investigations take this requirement into account, storing substantial amounts of semantic information in the computer memory, or presenting a visual display of the scene to which the language refers, or both.

"Understanding" language can mean different things in different contexts. If the language gives task instructions, then understanding is tested by whether the system undertakes to perform the task. If the language is descriptive or informative, understanding is tested by whether the system can store a representation of the information in its memory.

Not all of the language understanding programs that have been written purport to process text in the same way that humans do; many of them are exercises in artificial intelligence rather than psychological investigations. Nevertheless, a great deal has been learned about the requirements that any system must meet in order to understand language, and a substantial amount has been learned, too, about how people meet these requirements.

Lifelike problem domains

When a manager makes a decision about inventories or a new product line, he is not merely exercising his problem-solving

---

[21] A number of examples are described in Simon & Siklóssy, eds., *Representation*, and others are cited in the references there.

capabilities. He is also drawing upon a whole body of knowledge he possesses about business methods in general, about the industry in which his firm operates, and about his firm in particular. In contrast with the puzzlelike domains that were used in much of the early research on problem solving, the domains of real-life problems have substantial semantic content.

There is no trick in storing large amounts of information in a computer memory. Every automated accounting system or airlines reservation system does that. It is another matter to store information about a whole range of different matters in such a way that the system will notice what particular information might be relevant to the current situation, and will be able to retrieve it. A number of schemes have been constructed in the past five or ten years for simulating the organization of human long-term memory, and the associative processes used to retrieve information from that memory.[22]

An early, and quite successful, attempt to simulate human problem solving in an information-laden situation was Geoffrey Clarkson's model of the decision processes of a bank trust officer making investments in stock portfolios.[23] This program drew upon a memory that contained lists of industries each of which had a list of companies associated with it. This memory held economic and accounting information about the general economy, the industries, and the individual companies. The information was used first to select a basic list of stocks that were suitable for investment, and then to use this list, in combination with information about the objectives of each trust fund, to make the investment decisions for the individual funds. Clarkson also showed, by taking thinking-aloud protocols of a trust officer making decisions, that the processes in his program were in close agreement with the human processes.

A number of studies of problem solving in information-rich domains are under way at the present time.

---

[22] A survey of the recent work on the organization of semantic memory will be found in John Anderson & Gordon Bower, *Human Associative Memory* (Washington: V. H. Winston, 1974).

[23] This program is reported in Edward A. Feigenbaum & Julian Feldman, eds., *Computers and Thought* (New York: McGraw-Hill, 1963).

Where Does Simulation of Problem Solving Lead?

Success in simulating human problem solving can have two kinds of consequences: It may lead to the automation of some organizational problem-solving tasks; it may also provide us with means for improving substantially the effectiveness of humans in performing such tasks. Let us consider these two possibilities and their interrelations.

automation of nonprogramed decision making

If I am right in my optimistic prediction that we are rapidly dissolving the mists that surround nonprogramed decision making, then the question of how far that decision making shall be automated ceases to be a technological question and becomes an economic question. Technologically, it is feasible today to get all our energy directly from the sun, and to be entirely independent of oil, coal, or nuclear fuels. Economically, of course, it is not feasible at all. The capital investment required for direct conversion of the sun's rays to heat is so large that only under special circumstances and for special applications is the process efficient even at the present high fuel costs.

Similarly, the fact that a computer can do something a person can do does not mean that we will employ the computer instead of the person. Computers are today demonstrably more economical than people for most large-scale arithmetic computations. In most business data-processing tasks they are somewhere near the break-even point—whether they can prove themselves in terms of costs for any particular application depends on the volume of work and on the biases of the individual who makes the calculations.

To put the matter crudely, if a computer rents for $10,000 a month, we cannot afford to use it for nonprogramed decision making unless its output of such decisions is equivalent to that of about five persons at middle-management levels. Our experience to date—which is admittedly limited—suggests that computers do not have anything like the comparative advantage in efficiency over humans in the area of heuristic problem solving that they have in arithmetic and scientific computing.

There is little point in a further listing of pros and cons. As computer design evolves and as the science of programing continues to develop, the economics of heuristic problem solving by computer is changing rapidly. As it changes, we have to reassess continually which tasks are better automated and which tasks are better put in the hands—and heads—of the human members of organizations. About the only conclusion we can state with certainty is that the boundary between human and computer in data-processing work has moved considerably each five years since the initial introduction of computers into business, and will almost surely continue to move at a comparable pace.

improving human decision making

We should allow neither our anxiety nor our fascination to divert all our attention to the developments in automation. At least as important are the possibilities that are opened up for improving human problem-solving, thinking, and decision-making activity.

This is not the first time that the human species has made innovations that greatly improved its own thinking processes. One such innovation was the discovery of writing—the greatest importance of which lay perhaps in the aid it gave to immediate memory in performing such tasks as multiplying numbers together. (Put away your pencil and scratch pad and try to multiply two four-digit numbers.) A second such innovation was the discovery of the Arabic number system with its zero and positional notation. A third was the invention of analytic geometry and the calculus, permitting problems of scientific theory to be considered and solved that were literally unthinkable without these tools.

All of these aids to human thinking, and many others, were devised without understanding the process they aided—the thought process itself. The situation before us now is that we are acquiring a considerable understanding of that process. We shall soon be able to diagnose with accuracy the difficulties of a specific problem solver or decision maker in a specific domain, and we shall be able to help him modify his problem-solving strategies in specific ways. We shall be able to specify exactly what it is that a man has to learn about a particular subject—what he has to notice, how he has to proceed—in order to solve effectively problems that relate to that subject.

We have little experience as yet for judging how much improvement in human decision making we might expect from the application of this new and growing knowledge about thought processes to the practice of teaching and supervision and to the design of organizations. Nonetheless, we have reason, I think, to be sanguine at the prospect.

Conclusion

In this chapter we have contrasted the traditional ways that organizations have had for making programed and nonprogramed decisions with the new techniques that first began to appear after the Second World War, and whose appearance and diffusion were greatly accelerated by the invention and introduction into business organizations of the modern electronic computer.

The traditional methods for making routine, programed decisions have already been revolutionized by the invention and wide application of the new mathematical techniques that go under the names of operations research and management science. There are few business or governmental organizations of any size whose operations have not been significantly affected by these techniques, which first began to find application in the early 1950s. Although logically independent of the computer, these tools commonly require large amounts of computation for their application, and a large part of their practical significance for management must be attributed to the coincidental appearance of electronic computers at just the time when the OR techniques called for them.

The traditional methods for making nonprogramed decisions in organizations—involving large amounts of human judgment, insight, and intuition—have not undergone any comparable revolution. At the present time, however, we can see the makings of such a revolution in the basic research on heuristic problem solving and the simulation of human thinking that has been carried out in the past twenty years. We now know a great deal about what goes on in the human head when a person is exercising judgment or having an intuition, to the point where many of these processes can be simulated on a computer. In this chapter, I have tried to provide a brief description of this research and the picture of human thought processes it has produced.

Our growing understanding of nonprogramed decision making will bring about two quite different kinds of change in management. On the one hand, it will open up prospects for automating certain aspects of the decision-making process in the nonprogramed domain, just as operations research has permitted the automation of many aspects of programed decision making. On the other hand, by giving us deep insights into human thought processes, it will provide new opportunities, especially through education and training, to improve the capabilities of human beings in general, and executives in particular, for making decisions about difficult, ill-structured, complex situations.

The rest of this book is devoted to examining the implications of these developments for organizations and for the people in them. The next chapter will discuss their impact on work and people's feelings toward work, in both blue-collar and white-collar jobs. The fourth chapter will analyze implications for the shape and design of organizations, and for the jobs of managers in them. The fifth chapter will consider the broader implications of computers, automation, and the growth of technology as a whole for our economy and for our society.

3

the impact
of computers
on the workplace

The impact of computers and automation upon managers must be examined in the broader context of their impact upon the factory and the office, the setting within which managerial work takes place. There is a myth abroad that the latter impact is very great, and that it is almost all bad. It is the purpose of this chapter to test the myth against the available facts.

In its starkest outlines, the myth runs somewhat as follows: Computers and automation bring about the dehumanization of work, and dehumanization, in turn, causes alienation from work and from society. This bare outline can be fleshed out in a variety of ways, not all of them mutually consistent. In one scenario, the introduction of the computer brings about a high degree of centralization in organizations and subjects employees ever more heavily to the yoke of authority. The authoritarian atmosphere of organizations is dehumanizing for those who have to spend their lives in them and causes their alienation. This is the scenario that has been proposed by adherents of the human relations approach to administration.

A second, more flamboyant, scenario is Toffler's thesis of "future shock."[1] Computers and automation, along with other innovations in technology, according to Toffler, are bringing about such rapid change in our society that people are unable to cope psychologically with the continual bombardment of new information and with the repeated alterations and disruptions of their lives. This scenario was vividly enough portrayed not only to achieve best-seller status, but even to worry sufficiently a President of the United States so that he asked members of his Science Advisory Committee, in 1971, for an evaluation of it.

A third scenario, perhaps the most widely accepted of all, is that computers and automation, by routinizing jobs, introduce tedium into the workplace, thereby dehumanizing it and producing the consequent alienation.

There are other ills attributed to computers and automation as well. There is the concern that automation will create permanent unemployment of much of the workforce. There is the concern that continuing increases in productivity will create a glut of goods and services. Finally, there is the concern that continually increasing productivity cannot be sustained in the face of the limits of resources and of the capacity of the environment to absorb pollutants. Consideration of these topics will be postponed to Chapter 5, where we consider the broader economic and technological impacts of automation. This chapter will focus on the dehumanization-alienation hypothesis.

One way in which we can go about our inquiry is to work backward from the presumptive end result—alienation—to its causes, looking at such empirical evidence as is available at each step in the analysis. First, how widespread is alienation, and what are its trends? Second, to what extent can the alienation one sees in our society be traced back to the dehumanization of work or the disorienting impact of rapid technological change? Third, does increasing computerization and automation de-

[1] Alvin Toffler, *Future Shock* (N.Y.: Random House, 1970).

humanize work, either by making it more tedious or by subjecting workers to demeaning authority relations?

How Widespread Is Alienation?

The term *alienation* is widely used today to label a whole range of dissatisfactions with society, particularly those dissatisfactions that lead to distrust of our social and political institutions and withdrawal from participation in them. Neither the term nor the phenomenon is new.

alienation from work

The concept of alienation in its present form finds its origins in the writings of Karl Marx, particularly his early philosophical writings that were only published in the first half of this century. But even in the *Communist Manifesto* (1848) we find a succinct statement of it:

> Owing to the extensive use of machinery and to the division of labor, the work of the proletarians has lost all individual character, and consequently, all charm for the workman. He becomes an appendage of the machine, and it is only the most simple, most monotonous, and most easily acquired knack that is required of him.

Alienation, then, is no special product of our age and time; it was alleged to be the common state of the workingman nearly 130 years ago. Unfortunately, we did not have attitude surveys and public opinion polls at that time that would enable us to compare levels of alienation then and now. We have had job satisfaction surveys in individual firms for the past forty years, but comparable ones of a sample of the national population only for the past fifteen. Still, this covers the period of large-scale introduction of computers into industry.

A careful review and reanalysis of the two main sources of job satisfaction data—seven surveys carried out by the Survey

Research Centers of the Universities of Michigan and California and the National Opinion Research Center, from 1958 to 1973, and eight Gallup polls, from 1963 to 1973—was undertaken in 1974 under a grant from the U.S. Department of Labor.[2] That study concluded (page 1):

> In spite of public speculation to the contrary, there is no conclusive evidence of a widespread, dramatic decline in job satisfaction. Reanalysis of 15 national surveys conducted since 1958 indicates that there has not been any significant decrease in overall levels of job satisfaction over the last decade.

The words "no dramatic decline" and "no significant decrease" in this conclusion are conservative, for the data upon which the conclusions are based show no decline whatsoever. While the report does not provide any evidence on trends over a longer period—and, indeed, little evidence is available—it does caution us against accepting at face level claims about such trends that are based upon subjective impressions instead of hard facts.

The studies that have been made also do not tell us much about the *absolute* levels of satisfaction. In general, in these surveys, 80 to 90 percent of the workers polled reported themselves at least "moderately satisfied" with their jobs. This reply might be compatible with a considerable amount of alienation: It could reflect apathy or a reluctance to admit career failure. Lower levels of satisfaction do appear when workers are asked whether they would select the same career again, or whether they would choose their present jobs over all other jobs. But to interpret negative answers to the latter questions as "alienation" is to stretch the meaning of that term.

Do we think that 100 or 150 years ago a larger percentage of workmen (including farmers) would have said they were

[2] *Job Satisfaction: Is There a Trend?*, Manpower Research Monograph No. 30, U.S. Department of Labor, Washington, D.C., U.S. Government Printing Office, 1974. For a detailed analysis of these data, see Robert P. Quinn and Linda J. Shepard, *The 1972–73 Quality of Employment Survey*, Ann Arbor: Survey Research Center, Institute for Social Research, University of Michigan, 1974.

"moderately satisfied" with their jobs? In the absence of data, each reader will have to make up his mind on that question. For myself, I must return the Scottish verdict: "Not proven."

other forms of alienation

There are other forms of alienation besides job dissatisfaction. We can also try to assess people's satisfaction with their society, or with the quality of life. We can gauge what degree of trust and confidence they feel in political and other social institutions. Here there is some evidence of short-term (i.e., in the past decade) decline in assessments of the quality of life and in trust in political and economic institutions. Answers to the question, "Are you happy?" have shown no comparable trends. (There was perhaps a slight increase in reported happiness from the late forties to the late fifties or early sixties, and then a slight decline.)

How shall we interpret all of these trends or absences of trends? With respect to the general stability of the indices, there is a fairly straightforward explanation. When you ask someone, "How happy are you?" or "How satisfied are you with your job?", before he can answer he must ask himself, explicitly or implicitly, "as compared with what?" Satisfaction and happiness are measured relative to a level of aspiration. And the evidence from psychological research indicates pretty clearly that aspirations adjust themselves to realities and to alternatives. They probably adjust upward more readily than they adjust downward, but by and large what people aspire to is closely related to what they might reasonably expect to attain. Their expectations, in turn, are based on past experiences and on their observations of others around them. To put the matter in its simplest terms, if satisfaction means being about as well off as most other people, then most people will be satisfied most of the time. And so, it appears, they are. Short-term changes in the average level of satisfaction may be expected if sudden changes occur in social conditions, but we would expect the average to return gradually to its equilibrium level.

This interpretation of the meaning of self-assessments of happiness in terms of aspiration levels is supported by data from international comparisons of personal happiness ratings.[3] Within any single country there is a substantial correlation between happiness ratings and family income. Evidently the rich are aware of the existence of the poor, and vice versa, and people compare with others who are better or worse off economically than themselves in evaluating their own happiness. In comparisons between countries, however, there is almost no relation between the ranking of a country by average happiness and its ranking by per capita gross national product. The comparisons do not appear to extend across international boundaries; people judge their happiness in relation to average levels in their own society.

It is true that the average happiness levels reported in some societies are consistently higher than those reported in others; but since these differences are uncorrelated with average incomes, they must be attributed to "national character" or to differences in the nuances of the words for "happiness" in different languages rather than to differences in "objective"—i.e., economic—circumstances.

Even measures of attitudes like "trust" undergo adjustments of their zero points. In a society in which, in the past, unlocked bicycles were seldom stolen, a rash of thefts would lead to a short-term fall in feelings of trust, but a longer-term increase in the sale of bicycle locks. We "trust" people if they do not behave in ways that are both deleterious to us and unexpected. We do not mistrust people if they are selfish, aggressive, or competitive in expected ways and to expected degrees.

Attitudes toward matters in which people have direct experience are considerably more stable than attitudes toward matters on which they are informed only at second hand. Over the past decade, the public media have tended to stress aspects of our society that might be expected to reduce satisfaction and trust: the attacks of the New Left, crime and violence, the Vietnam War, the environmental and energy crises, unemployment, and Watergate. This abundance of bad news has not persuaded people that they are less happy in their personal lives—in contra-

[3] These findings have been summarized recently in Richard E. Easterlin, "Does Economic Growth Improve the Human Lot? Some Empirical Evidence," pp. 89-125 in *Nations and Households in Economic Growth*, Paul A. David and Melvin W. Reder, eds. (New York: Academic Press, 1974).

diction to their own direct assessments of those lives—but has changed attitudes in such areas as trust of government, where their information is obtained largely from these same cheerless media.

There is much more that could be said on this subject, but it would take us away from our main task: to determine to what extent, if at all, the introduction of automation and computers has produced or will produce substantial alienation of people from their jobs or from society. We have only very recent statistics on these matters, but the testimony of literature from classical times to the present suggests that alienation is no new phenomenon in human societies. Nor is there any real reason to think that there is more alienation in our present society than in the societies that have preceded it over the past several hundred years.

Dehumanization of Work

Since we have not found any evidence of increased alienation, it might seem that we do not need to inquire whether computers and automation are contributing to the dehumanization of work. However, in the light of what has been said about adjustable aspiration levels, it could be that work was indeed being dehumanized, but that people were adjusting to the change. We might not like to see such a change take place even if people were not expressing dissatisfaction with it.

In discussing possible dehumanizing (or humanizing) effects of computers and automation, we must consider two kinds of effects: direct effects on the workplace where the computers or automation are introduced, and indirect effects produced by resulting changes in the profile of occupations.

direct effects of computers and automation

Sizable differences have been found in worker satisfaction among blue-collar workers in different kinds of factories. The important variables would seem to be job variety and worker control over timing. In the imaginations of persons unfamiliar with factories, the usual picture of a mechanized factory is a textile spinning mill or an automobile assembly line. In operations like these, tasks are highly routinized and the operatives must adapt their pace to the pace of the machines. To the extent that there is blue-collar alienation from work, it tends to concentrate in these kinds of workplaces. Less mechanized large-scale hand assembly operations paid on a piecework basis tend also to be low in satisfaction.

Of course factories with these characteristics are not typical of modern automation and computerization. They represent old-fashioned modes of operation, which were perhaps common enough when Charlie Chaplin made the movie *Modern Times,* but which are now gradually disappearing. As work becomes more fully automated, the worker ceases to be a direct link in the production process. As long as the process is operating normally, the machines turn out the physical product without the intervention of human hands. The operative may have responsibilities for setting up jobs, monitoring the process to see that it is going properly, and intervening to handle difficulties. He is usually not tied to a specific physical location; he may or may not have frequent opportunities for interaction with other human beings. This kind of automated operation is not completely new: one could have seen something like it in a central electrical generating station fifty years ago. It is more and more becoming the typical form of the organization of work, not only in factories but in large-scale computerized clerical operations as well.

Forms of mechanization

Empirical studies of the psychological and social effects of factory organization and technology generally indicate that the newer forms of automation provide less stressful and alienating work environments than do the older forms of mechanization. Blauner, for example,[4] studied four industries in depth—printing, a traditional craft industry; textiles, a machine-tending industry; an automobile assembly line, highly mechanized with highly specialized jobs; and the highly automated continuous-process chemical manufacturing industry. He interpreted alienation to

[4] Robert Blauner, *Alienation and Freedom: The Factory Worker and His Industry* (Chicago: University of Chicago Press, 1964).

mean feelings of powerlessness, meaninglessnes, isolation and self-estrangement. He found few indications of alienation in printing (craft technology) and chemicals (highly automated); considerably more in textiles (machine-tending); and most of all in automobile assembly. The critical factors that determined degree of alienation appeared to be less the technology itself than such factors as degree of control over one's work environment, responsibility for product, the social structures of communities from which workers were drawn, and opportunities for stable employment and advancement. Nevertheless, it is notable that the industry that best typifies modern automation—chemicals—was substantially less alienating than the two that typify the older kinds of mechanization—textiles and automobile assembly.

This analysis of the effects of factory automation upon feelings of alienation may help us to understand and predict the likely effects of office automation, which is a more recent phenomenon. We can ask to what extent office automation shares the characteristics of assembly line or machine tending jobs, and to what extent it more nearly resembles the automation of a continuous process plant. The tasks in a computerized office that most resemble the older mechanization are the jobs of preparing machine-readable source documents and entering them into the computer. These jobs do not play a larger role in computerized installations than in those using older kinds of business machines, but may be somewhat changed in character, as we shall see. The jobs that most resemble the more modern kinds of automation are those involved in programing and operating the central computer itself. As the computer technology advances, more and more of the information to be input to the computer is machine readable, requires less and less human intervention to transfer it to the machine. Hence we would expect the first class of jobs to have a declining part in the total operation as compared with the second class.

The introduction of computers

In the late 1960s, there was a spate of empirical case studies of the effects upon work of the introduction of computers into the office. Most of these studies, unfortunately, were made at the time when automation was taking place or shortly thereafter, making it difficult to separate out transient effects, of sorts that might result whenever large-scale changes of any kind are introduced into an organization, from effects that might be expected to be more or less permanent. Further, in the analysis of the findings of these studies, immediate direct effects at the work place have not always been separated from net effects after secondary adjustments have taken place elsewhere in the economy. These problems, together with the fact that the authors of some of the studies seem to have approached them with rather definite expectations of what they would find, mean that the data and their interpretation must be scrutinized with care.[5]

Whisler, who studied about twenty companies in the insurance industry, sums up his findings about the impact of computers on job content as follows (*Impact,* page 131):

> With respect to impacts on jobs, we may conclude that, on balance, clerical jobs have become more routinized while supervisory jobs have tended toward enlargement. Yet, no clear trends are visible in terms of the routinization or enlargement of managerial jobs. Moreover, where computer systems have been installed and refined over a considerable period of time—chiefly at the clerical and supervisory levels—the result has been to diminish the level of interpersonal communication; and, where computer systems are in the process of development—chiefly at the managerial level, occasionally at the supervisory level—the result is to increase the flow of interpersonal communication. Skills have been affected at all organizational levels, but skill changes are most pervasive at the clerical level, their incidence diminishing steadily at successively higher levels. Overwhelm-

[5] Four books will give the reader a good picture of this literature: Ida Russakoff Hoos, *Automation in the Office* (Washington, D.C.: Public Affairs Press, 1961); H. A. Rhee, *Office Automation in Social Perspective* (Oxford: Basil Blackwell, 1968); Thomas L. Whisler, *Information Technology and Organizational Change* (Belmont, Cal.: Wadsworth, 1970); and Thomas L. Whisler, *The Impact of Computers on Organizations* (New York: Praeger, 1970). Although there is general consistency in the empirical findings reported by all four books, there are substantial differences in the conclusions drawn by the authors. Whisler's books, coming later than the others, suffer the least from the confusion of transient with permanent effects, and are the most objective and detailed in their presentation of the raw data. In my analysis here, I have generally followed Whisler's interpretations, which seem to me the ones best supported by the evidence.

ingly, the effect is perceived to be an upgrading of skills, though the upgrading effect is tempered at the clerical level, with about a third of the changes in the direction of downgrading.

Whisler, noting the apparent contradiction in the finding that clerical jobs were routinized while clerical skills were upgraded, sought further clarification from some of his respondents. The explanation given was that the new jobs placed greater demands on the employees for accuracy and reliability in performance. They now had to meet closer specifications in their work, and needed greater skill to meet them. Whisler's study did not determine whether the jobs were now perceived as being more or less pleasant or more or less boring than they had been previously.

What is striking about all of the studies reported in the literature is that, except in instances where the rapid and poorly planned introduction of automation caused significant transient disturbances, the effects upon the nature of work were difficult to detect, and were far from consistent in direction. Whisler found, for example, that among the clerical jobs in the offices he studied, 31 percent of the jobs were routinized, while 22 percent were enlarged and 47 percent were unchanged (*Impact*, page 131). Six companies reported that clerks now communicate less with other clerks; eight reported that communication was now greater or the same (*ibid.*, page 136). Ninety percent of computer-affected departments reported that changes had occurred in clerical skills (*ibid.*, page 140). Of these, 70 percent reported that skill requirements had been raised, 30 percent that they had been lowered.

If large effects occurred from job routinization, one would expect to see them most clearly in jobs like those of keypunch operators, but the evidence here is highly variable. Rhee (*Office Automation*, page 125) observes:

> Although one departmental manager described the work as "extremely boring, always shift work, with no intelligence looked for," it is reported in another inquiry that "the peripheral equipment operators felt that the level of difficulty of their present assignment represented an increase over their previous jobs in the Hollerith department. . . ." On the point of career aspirations

of the key punch operators, findings in various case studies differ widely.

In summary, there is nothing in the evidence to suggest that office automation has changed or will change the nature of work in a manner conducive to large-scale alienation. It is significant that most of the studies of the impact of office automation were carried out in the middle sixties, when the technology was new and its possible impact hard to anticipate. Today, when automation is far more widespread than it was at the time the studies were made, we hear very little about its effects. Perhaps the reason is that, once the changes had been introduced and initial disturbances had had time to work themselves out, there was not much to study.

Transient effects of automation

Since the direct effects of office automation upon jobs appear to be of only modest significance, we turn to the indirect effects. Before taking up that topic, however, some final words are needed on what I have been calling *transient effects*. It is well known that proposed new changes in organizations can meet with enthusiastic support or violent resistance. The key to their reception is whether they are perceived as being done *by* the people involved, or *on* those people. No one likes to be subjected to change in which he has no active part; most people are exhilarated by being agents of change. The perception of whether one is victim or agent of change depends, in turn, mainly on two factors: (1) whether the change will have desirable or undesirable personal consequences, and (2) the extent to which one is informed of the coming change and has an opportunity to participate in preparing for it and shaping it.

We have probably learned something in the past fifty years about how to introduce changes into organizations with careful regard for their effects on people's lives and without causing major turmoil and resistance. We have not always practiced what we have learned. The cases reported in the literature where office automation has caused major upset are examples of mismanagement of change, and not of inevitable undesirable human effects of automation.

indirect effects of office automation

Factory and office automation reduce costs by reducing the labor required to perform a given amount of work. They are labor-saving technologies. The jobs they eliminate are mostly jobs that were already relatively routine. Therefore, when we look at the impact on the labor force as a whole, we would expect to see automation bringing about an overall decrease in the percentage of persons engaged in routine jobs of these kinds. The missing employees will not be unemployed (see Chapter 5, below), but will appear in different kinds of jobs in the new equilibrium situation. There will be, for example, a larger percentage of employees than before in service occupations, and probably also in technical occupations.

In order to see what this shift implies, let us return to the work satisfaction studies discussed earlier.[6] These studies reveal substantial differences among broad occupational groups in average satisfaction. As might be expected, occupations rank from high to low in satisfaction in the following order: professional and technical; managers, officials and proprietors; sales; craftsmen and foremen; service workers, except household workers; clerical; operatives; and nonfarm laborers. (The samples were not large enough to permit ranking farmers, farm laborers or private household workers.)

Operatives and nonfarm laborers were only about one half of a standard deviation below the mean in satisfaction, and professional and technical personnel only about a third of a standard deviation above the mean, so the differences were not enormous. Nevertheless, it is clear that if factory operatives and clerical workers decline as a fraction of the labor force, while service workers, sales personnel, and professional and technical workers increase, there will be a net increase in reported job satisfaction —unless, of course, a compensating shift takes place in aspiration levels, a possibility we must not dismiss.

In the light of these presumptive long-term trends in the composition of the work force, it should not be surprising that

[6] *Job Satisfaction,* Manpower Research Monograph No. 30, p. 10.

there has been no downward trend in average job satisfaction. In fact, we might ask why there has not been a trend upward. There are at least two reasons for this. One is the motility of aspiration levels, which has already been discussed. The other is that the shift in the occupational profile over ten or fifteen years has not been large enough to cause a visible change in the gross statistics of job satisfaction.

Are Organizations Authoritarian?

Thus far we have been examining the series of links that might connect changes in work brought about by the new decision-making tools with alienation, via the mechanism of job routinization and consequent loss of job satisfaction. We need also to consider a second set of mechanisms that could produce (or reduce) alienation, in this case through changes in authority relations. In the next chapter, we will consider in detail how authority relations might be altered by changes in decision-making processes. In preparation for that discussion, we need some insight into the possible psychological significance of such changes.

the nature of authority

The exercise of authority is one of the most pervasive phenomena of life in organizations. Authority is exercised whenever a person allows his decisions to be guided by decision premises provided to him by some other person (or, for that matter, by a computer). Since, in modern organizations, there is a great deal of division of labor in the decision-making process as well as in action processes, decision premises are continually being generated in one organizational location for communication to other locations where they influence decisions. Each such instance is an exercise of authority.

In the modern human relations literature, there is a strong tendency to view authority relations with great suspicion, which sometimes leads to confusion between "authority" and "authoritarianism." [7] To avoid that confusion in what follows, I will use the term "authoritarian" only to refer to management that uses authority to an excessive degree. "Excessive" is, of course, also a word that will need specification.

is authority dehumanizing?

In the literature that is critical of the uses of authority, there are two lines of argument, both resting on similar premises about human motivation. The first is that acceptance of authority denies participation in the decision-making process. Participation in decision, however, is essential to achieve understanding and enthusiasm in carrying out decisions. Hence, increasing participation at the expense of authority will increase organizational effectiveness. Since the 1930s, when the famous Hawthorne experiments were carried out, there have been numerous empirical studies showing that, under some circumstances at least, broadened participation in decision making can increase the acceptance of organizational decisions, particularly decisions to change the organization in major ways. [8]

The second argument takes as its starting point a theory of motivation advanced by Abraham Maslow, which postulates that human needs can be arranged in a hierarchy, from basic physiological needs at the bottom to the need for self-actualization at the top. [9] The lower needs take precedence over the higher, and only as the former are satisfied do the latter come into play. Subjection to authority, it is then asserted, is inimical to self-actualization. Organizations stifle man's highest drives, causing him to regress to a state of dependency that leaves no room for the exercise of his imagination and his creative fulfillment as a person.

The argument is appealing, but speculative. It rests on the dubious assumption that if something is good—in this case, free-

[7] See my "Organizational Man: Rational or Self-Actualizing," in *Public Administration Review*, 33:346-53, July-August 1973.

[8] Gary Dressler, *Organization and Management* (Englewood Cliffs, N.J.: Prentice-Hall, Inc., 1976), pp. 35-37.

[9] Dressler, *Organization*, pp. 208-10.

dom from subjection to excessive authority—then more of it is better. But both laboratory experiments and everyday observation suggest that human beings are not at their most creative and self-fulfilling in the least constraining environments. Games, to which human beings turn as an important means of self-expression, impose highly constraining sets of rules upon behavior, and games are not usually improved simply by relaxing their rules.

In fact, human beings are distinctly uncomfortable when they are placed in environments so lacking in structure that each decision is a major intellectual task, and no clues are provided as to what to do next. Routine is a welcome refuge from the trackless forests of unfamiliar problem spaces.[10] People find the most interest in situations that are neither completely strange nor entirely known—where there is novelty to be explored, but where similarities and programs remembered from past experience help guide the exploration.

Nor does creativity flourish in completely unstructured situations. The almost unanimous testimony of creative artists and scientists is that their first task is to impose limits on the situation if the limits are not already given. The pleasure that the good professional experiences in his work is not simply a pleasure in handling difficult matters; it is a pleasure in using skillfully a well-stocked kit of well-designed tools to handle problems that are comprehensible in their deep structure but unfamiliar in their detail.

If this is so, then organizations will be most human, humane, and conducive to self-actualization when they strike a balance between freedom and constraint. This principle by itself does not provide a very helpful guide to organizational design, because it does not tell us what is a "balance." However, it is characteristic of all sorts of design, architectural included, that the best structures are not produced by maximizing along a single dimension, but by seeking some middle ground that accommodates a whole assemblage of criteria and desiderata.

Individuals are likely to feel most comfortable in orga-

[10] For a careful examination of the psychological evidence on this topic, see D. E. Berlyne, *Conflict, Arousal, and Curiosity* (New York: McGraw-Hill, 1960).

nizations when they are given decision-making tasks that are challenging, but in which they feel competent—in which the framework of decision provides definite points of structure that their problem-solving tools can lay hold of. Fortunately, these are also the conditions under which they are likely to be able to make good decisions. The decision premises they receive from other parts of the organization need not be perceived as subjecting them to a foreign authority, but can be viewed as linking them to broader organization goals and providing them with the expertise of other organization members. Participation in an authority structure need not be constrictive, it can also be freeing.[11]

trends in authority relations

There has been a large and continuing change in the nature of the authority relations in our society, not only in formal organizations, but also in the relations between parents and children, between teachers and students, between husbands and wives, between ministers and congregations—the list could go on almost indefinitely. Most of us (because we are products of our own culture, if for no other reason) view these changes as beneficial. The older relations now seem to us authoritarian, as they certainly are relative to the present scene.

If we think the consequences of this change have, on balance, been desirable, it does not follow that the consequences of an indefinite continuation of the trend are equally desirable, or desirable at all. An excessive preoccupation with authority relations is often symptomatic of excessive needs for power and control. And thinking too much of organizational life as a struggle for power leads to thinking too little of it as a means of expressing man's creative impulses and meeting his needs. This is hardly a novel observation: It has been remarked before that demands for freedom *from* control are frequently transformed into demands for freedom *to* control. The desire to destroy authority relations often conceals the desire to replace them with one's own authority.

[11] This point is developed at greater length in my *Administrative Behavior*, 3rd ed. (New York: The Free Press, 1976), pp. 100-108.

Whether the adoption of new decision-making tools will lead to a greater or lesser use of authority in organizations will be considered in the next chapter. The present discussion warns us that, whatever our conclusion from that analysis, we cannot judge organizational designs merely by how much participation in decision making they provide for members. That is a simplistic criterion, only one among a great many that must be kept in view. Organizational authority is a useful, and even indispensable, tool for permitting human beings to cooperate toward common goals. The problem is to use it in such a way that it will generally be experienced as a source of supportive structure rather than arbitrary constraint.

Do We Suffer From Future Shock?

Before we leave the topic of the human effects of automation, we should ask whether the very rapidity of change in our society, only part of it induced by the computer and by automation, is so great that it is exacting a heavy psychological toll from us. The phrase "future shock" that Toffler coined to denote the psychological stresses produced by rapid change is so felicitous that it is hard to believe we are not suffering from it. Again, however, the concern is not new:

> Constant revolutionizing of production, uninterrupted disturbance of all social conditions, everlasting uncertainty and agitation distinguish the bourgeois epoch from all earlier ones. All fixed, fast-frozen relations, with their train of ancient and venerable prejudices and opinions, are swept away, all new-formed ones become antiquated before they can ossify. All that is solid melts into air, all that is holy is profaned, and man is at last compelled to face with sober senses, his real conditions of life, and his relations with his kind.

That is future shock, vintage of 1848, again from the pages of the *Communist Manifesto*. But there is a difficulty with the argument, both in 1848 and in 1976. The former date is too early and the latter is too late. If there are limits on the rate of change to which human beings can adjust, then those limits should have been revealed most clearly to the two or three generations that had to undergo the change from a rural, agricultural, horse-powered society to an urban world with railroads, telegraph, steamships, electric light, automobiles, and finally the airplane—that is, the generations between the Civil War and the First World War. Nothing that has happened in my lifetime, with the possible exception of The Bomb, has so changed the basic terms of human existence as those new technologies did.

There is no evidence of which I am aware that human breakdown in the face of the stress of change reached higher levels during that period in which a rural world became an urban world than during the periods that preceded it or followed it. If it did not, then the hypothesis that we are suffering at the present time from anything that can be described as "future shock" is erroneous. If there is future shock in the world today, it is far more likely to be found in the developing nations, where the great transition is still taking place, than in the developed world, where it has had fifty or one hundred more years to be digested.

Even if there is no reason to believe that our generation lives under more stress than the generations that preceded it, it is sensible to ask what can be done to handle the stress that does exist. Stress in the face of social and technical change is more likely to be produced by insufficient cognitive structure than by an excess of structure. Hence, any measures we can take to improve the decision-making capabilities of ourselves and our organizations should be therapeutic in their effect.

Conclusion

In this chapter I have examined the ways in which the new decision-making technology and the computers and automation associated with it could have impact on the nature of work—and particularly blue-collar and clerical work—in modern organizations. There has been a great deal of nervousness, and some prophetic gloom, about human work in highly automated organizations. An examination of such empirical evidence, and an analysis of the arguments that have been advanced for a major impact of automation upon the nature of work has led us to a largely negative result.

There is little evidence for the thesis that job satisfaction has declined in recent years, or that the alienation of workers has increased. Hence, such trends, being nonexistent, cannot be attributed to automation, past or prospective. Trends toward lower trust in government and other social institutions flow from quite different causes.

An examination of the actual changes that have taken place in clerical jobs as the result of introducing computers indicates that these changes have been modest in magnitude and mixed in direction. The surest consequence of factory and office automation is that it is shifting the composition of the labor force away from those occupations in which average job satisfaction has been lowest, toward occupations in which it has been higher.

The argument that organizations are becoming more authoritarian and are stifling human creativity flies in the face of long-term trends in our society toward the weakening of authority relations. Moreover, the psychological premises on which the argument rests are suspect. Far more plausible is the thesis that human beings perform best, most creatively, and with greatest comfort in environments that provide them with some intermediate amount of structure, including the structure that derives from involvement in authority relations. Just where the golden mean lies is hard to say, but there is no evidence that we are drifting farther from it.

Finally, while we certainly live in a world that is subject to continuing change, there is reason to believe that the changes we are undergoing are psychologically no more stressful, and perhaps even less, stressful, than those that our parents and grandparents experienced.

I do not draw the conclusion from all of these arguments that we live in the best of all possible worlds. The conclusion I do draw is that the human consequences we may expect from factory and office automation are relatively modest in magnitude, that they will come about gradually, and that they will bring us both disadvantages and advantages—with the latter possibly outweighing the former.

With this analysis of effects at the operative levels of organization concluded, it is time to look at the changes we may expect in organization structure and in the jobs of managers. This is the topic of the next chapter.

postscript

After I had completed this chapter, I came upon the eloquent recent book by Harry Braverman, *Labor and Monopoly Capital* (New York: Monthly Review Press, 1974), which reaches conclusions on alienation almost diametrically opposed to mine. (Braverman gave his book the subtitle, "The Degradation of Work in the Twentieth Century.")

There is almost no disagreement between Braverman and me about the effects of mechanization and automation upon skill requirements and the occupation and skill profile of the working population, but nearly total disagreement about the implications of job characteristics for worker alienation. It is perhaps significant that, in reaching his conclusions, Braverman does not refer to the findings of Whisler, and only casually and critically to those of Blauner, while he quotes at length Ida Hoos (who looked only at short-range phenomena during the changeover to office automation). Braverman also makes no use of the data from opinion polls on worker satisfaction and happiness. Instead, his whole argument rests on the "obvious" premise that work is degrading if it does not make complex cognitive demands on the worker.

Although I found Braverman's rhetoric powerful, I finished *Labor and Monopoly Capital* unconvinced that my chapter needed revising. Braverman's book introduces no empirical evidence that I had missed or ignored, and omits several crucial bodies of evidence, mentioned above, that seem to be critically relevant. Nevertheless, I recommend the book to anyone who wishes to hear the other side of the story stated forcefully.

Another recent publication on the history of attitudes toward work casts additional doubt upon Braverman's thesis. Alasdair Clayre, in *Work and Play* (New York: Harper & Row, 1974), has sought to determine whether there was a period prior to the Industrial Revolution when work was generally regarded as pleasurable and satisfying. He has analyzed not only the writings of philosophers and social commentators but also—and more important—the attitudes toward work expressed in folk literature and popular ballads. He finds few indications of such a Golden Age. By and large, daily work was apparently the same burdensome necessity for peasants and craftsmen as it is today for factory workers and clerks. Life's satisfaction and pleasure were mainly sought in leisure, not work.

4

organizational design:
man-machine systems
for decision making

Having examined the effects of computers and automation upon blue-collar and clerical work, we return in this chapter to a consideration of their impacts upon organization structure and the job of the manager. With operations research and electronic data processing we have acquired the technical capacity to auto-mate programed decision making and to bring into the pro-gramed area some important classes of decisions that were formerly unprogramed. Important innovations in decision-making processes in business have resulted from these discoveries.

With heuristic programing, we are acquiring the technical capacity to automate more and more areas of nonprogramed decision making. The coming decades will see many changes in business decision making and business organization that will stem from this second phase in the revolution of our information technology. In this chapter, I should like to explore how the world of business will be modified as both of these kinds of changes occur.[1]

Not all or, perhaps, most of the changes we may anticipate have to do with automation. As was pointed out in Chapter 2, the progress we may expect in the effectiveness of human decision-making processes is equally significant. In the present chapter, both advances in automation and advances in human decision making will figure, as will—perhaps most important of all—the integration of the human and electronic components of organizations into sophisticated man-machine systems.

The Information-Rich Environment [2]

The range of our inquiry cannot be restricted just to the computers that are to be found inside the organization's walls. In fact, it must not be limited to computers at all, but must take into account the whole modern communications technology of which they are a part. In particular, long-distance telephones and copying machines, and even the airplane, are important parts of the picture.

balancing information availability
with processing capacity

Abraham Lincoln, we are told, walked many miles to borrow books so that he could tap the world's information. He lived in a society in which information was scarce and dear. Any obtainable piece of it was to be treasured, and if possible, copied into human memory. The human information processor had adequate capacity to absorb most of the information that it could manage to lay its hands on.

[1] These developments have not lacked books of prophecy—including the first edition of this book and several essays that preceded it. Among recent writings, especially those few that pay attention to such empirical findings as are available, I have found Thomas L. Whisler, *Information Technology and Organizational Change* (Belmont, Cal.: Wadsworth, 1970), one of the most thoughtful and informative.

[2] A fuller discussion of this topic will be found in my *Administrative Behavior,* 3rd ed. (New York: The Free Press, 1976), Chapters 13 and 14.

Today, the problem for the human information processor, whether inside or outside an organization, is to select the communications to which he wishes to attend from the great flood of information by which he is buffeted. The whole concept of what it means to "know" something has been transformed. In the pre-computer era, a person knew something when he had it stored in his memory in such a form that he could retrieve it on appropriate cues. Or, if he had access to a good library, he could be said to "know" something if it was stored in one of the books shelved in the library, and if he could find it with the help of the catalog and other indexes, using his own problem-solving capabilities and memory store to help conduct the search.

Nowadays there are many additional ways of "knowing." A student or an engineer equipped with a hand-held calculator knows, in a real sense, the value of a trigonometric function that his calculator will compute in a second or two—about as fast as he could recover it from his own memory if it were stored there, and much faster than he could retrieve it from a table of functions. An airline reservation clerk has complete, up-to-the-minute knowledge of the bookings of flights, available to him from his computer terminal at the reservations desk. A manager of inventories has access to complete knowledge of the inventories and orders for thousands of items, as well as usage rates and ordering costs.

As these examples suggest, with the new technology there is often very little reason to transfer information to human memory, or even to man-readable memories like books, prior to the moment it is needed. The growing facility with which information stores can be accessed, or the information itself computed on demand, leads to holding information in "raw" or semi-processed forms, and especially in machine-readable rather than man-readable forms.

With contemporary technologies, memories may be widely distributed, and we must take a systems approach to knowledge instead of identifying "what is known" with what is stored in local memories. Generating information on demand or obtaining it from another part of the system are important alternatives to storing it—alternatives that are becoming available on more and more favorable terms of time and cost. If I have access to a telephone, for example, I "know" the names of the best American (or world) experts on any subject you care to mention. For by a succession of three or four telephone calls (following a sort of "Twenty Questions" strategy), I can locate the experts' names. With the usual defects of indexing, I could probably do no better if they were stored in my memory or my library.

Today the critical task is not to generate, store or distribute information but to filter it so that the processing demands on the components of the system, human and mechanical, will not far exceed their capacities. A good rule of thumb for a modern information system might be that no new component should be added to the system unless it is an information *compressor*—that is, unless it is designed to receive more information than it transmits. The scarce resource today is not information, but capacity to process it. Although the most visible components of our modern information systems are the printing, transmission and copying devices that enable them to pour out vast quantities of information, the crucial components are the sophisticated processors that can shield us from that torrent of symbols.

the new information technology

The main features of the information-processing technology of the near future (five to ten years, say) can be predicted with some confidence. Few innovations can be introduced widely in that short span of years unless the hardware and software required for them are already in existence. I would characterize this near-future technology as follows:

1. Substantially all information available to humans in verbal or symbolic form will also exist in computer-readable form. At present, it is costly to input the content of an existing book to a computing system. The book must be copied by a human typist or a photoscanning device. In the future, books and magazines will be stored in electronic memories at the same time that hard copy is produced for human use. The technology for doing this already exists, and is coming into increasing use. Many data that are now recorded or transcribed by humans will be transmitted directly to automated information-processing systems without human

intervention. For example, the quantity of product coming off an automated assembly line will be entered directly in the inventory records.

2. Memories in information-processing systems will be of sizes comparable to the largest memories now used by humans— for example, the book collection of the Library of Congress.

3. It will be feasible and economical to use natural language (English or another) in interrogating the memory of an information-processing system.

4. Any program or information that has proved useful in one information-processing system can be copied into another part of that same system or into another system at very low cost (and without severe problems of standardization).

5. Under the most optimistic assumptions about the power and capacity of prospective information-processing systems, they will remain puny in relation to the size of real-world planning problems. The chess board is a tiny microcosm compared with the macrocosm of real life. Nevertheless, there is absolutely no prospect of an increase in the power of information-processing systems to the point where they will play chess by "considering all possibilities" (some 10^{120})—much less to the point where they will carry out exhaustive searches of all the decision alternatives in real life situations.

Therefore, the significant limits on the power of information-processing systems for handling problems of planning and decision making in complex organizations will be limits on (1) knowledge of the laws that govern the systems being planned, (2) cleverness in discovering representations that handle salient characteristics of situations unencumbered by a mass of detail, (3) availability of powerful heuristic search methods of the kinds discussed in Chapter 2, as well as (4) availability of the relevant real-world data.

6. Information-processing systems will become increasingly capable of learning, in several senses of that term. In particular, they will be able to "grow" their own indexes as new information is added to their stores. Thus, the important contemporary bottleneck in human indexing and abstracting capability will become less significant as time goes on.

The discussion in this chapter of the shape of future organizations is based on these assumptions about the information technology, together with the conclusions reached in Chapter 2 about the decision-making technology that will be used with it.

The Hierarchic Structure of Organization [3]

An organization can be pictured as a three-layered cake. In the bottom layer we have the basic work processes—in a manufacturing organization, the processes that procure raw materials, manufacture the physical product, warehouse it, and ship it. In the middle layer we have the programed decision-making processes—the processes that govern the day-to-day operation of the manufacturing and distribution system. In the top layer we have the nonprogramed decision-making processes, the processes that are required to design and redesign the entire system, to provide it with its basic goals and objectives, and to monitor its performance.

Automation of data processing and decision making will not change this fundamental three-layer structure. It may, by bringing about a more explicit formal description of the entire system, make the relations among the layers clearer and more explicit.

Large organizations are not only layered, they are also almost universally hierarchic in structure. That is to say, they are divided into units which are subdivided into smaller units, which are, in turn, subdivided, and so on. They are also generally hierarchic in imposing on this system of successive partitionings a pyramidal authority structure. However, for the moment, I should like to consider the departmentalization rather than the authority structure.

Hierarchic subdivision is not peculiar to human organizations. It is common to virtually all complex systems of which we have knowledge. Complex biological organisms are made up of subsystems—digestive, circulatory, and so on. These subsystems are composed of organs, organs of tissues, tissues of cells. The

[3] For a more complete discussion of this topic, see Chapter 4, "The Architecture of Complexity," of my book, *The Sciences of the Artificial* (Cambridge, Mass.: M.I.T. Press, 1969).

cell is, in turn, a hierarchically organized unit with nucleus, cell wall, cytoplasm, and other subparts.

The complex systems of chemistry and physics reveal the same picture of wheels within wheels within wheels. A protein molecule—one of the organismic building blocks—is constructed out of simpler structures, the amino acids. The simplest molecules are composed of atoms, the atoms of so-called elementary particles. Even in cosmological structures, we find the same hierarchic pattern—galaxies, planetary systems, stars, and planets.

why complex systems are hierarchic

The near universality of hierarchy in the composition of complex systems suggests that there is something fundamental in this structural principle that goes beyond the peculiarities of human organization. There are at least three reasons why complex systems should generally be hierarchic:

1.   Among possible systems of a given size and complexity, hierarchic systems, composed of subsystems, are the most likely to appear through evolutionary processes. The mechanisms of natural selection will produce hierarchies much more rapidly than non-hierarchic systems of comparable size, because the components of hierarchies are themselves stable systems.

A metaphor will show why this is so. Suppose we have two watchmakers, each of whom is assembling watches of ten thousand parts. The watchmakers are interrupted, from time to time, by the telephone, and have to put down their work. Now watchmaker A finds that whenever he lays down a partially completed watch, it falls apart again, and when he returns to it, he has to start reassembling it from the beginning. Watchmaker B, however, has designed his watches in such a way that each watch is composed of ten subassemblies of one thousand parts each, the subassemblies being themselves stable components. The major subassemblies are composed, in turn of ten stable subassemblies of one hundred parts each, and so on. Clearly, if interruptions are at all frequent, watchmaker B will assemble a great many watches before watchmaker A is able to complete a single one.

The assembly of a large hierarchic system can proceed from

the bottom up, by successive mergers of subsystems as in the watchmaker metaphor, or from the top down, by successive splitting of units and the growth of the subunits. In both cases the resulting systems will be hierarchic.

2.   Among systems of a given size and complexity, hierarchic systems require much less information transmission among their parts than do other types of systems.

It was pointed out many years ago that as the number of members of an organization grows, the number of *pairs* of members grows with the square (and the number of possible subsets of members even more rapidly). If each member, in order to act effectively, has to know in detail what each other member is doing, the total amount of information that has to be transmitted in the organization will grow at least proportionately with the square of its size. However, if the organization is subdivided into units, it may be possible to arrange matters so that an individual needs detailed information only about the behavior of individuals in his own unit, and aggregative summary information about average behavior in other units. If this is so, and if the organization continues to subdivide into suborganizations by cell division as it grows in size, keeping the size of the lowest level subdivisions constant, the total amount of information that has to be transmitted will grow only slightly more than proportionately with size. Hence, the amount of communication required per organization member will remain nearly constant.

3.   With hierarchic structure, the complexity of an organization as viewed from any particular position within it becomes almost independent of its total size.

The complexity of an army, as viewed by its commanding general, or of the U.S. Steel Corporation as viewed by its president, is no greater than the complexity of a regiment or the complexity of a rolling mill, when these are viewed from the positions of those who manage them. A manager, no matter how large or small his total responsibility, interacts closely with a few subordinates, a few superiors, and a few coordinate managers. The number of persons with whom he communicates directly is approximately the same, whatever his level in the total organization.

With the rest of the organization he deals only in gross, aggregate terms, and mainly in indirect ways.

All levels of organization and organizations of all sizes have to be staffed with about the same kinds of human beings—allowing for a modest amount of selection for abilities as one moves upward. Since all these human beings have serial information-processing systems of roughly comparable power, large organizations can only operate if task complexity is nearly independent of organizational level. Hierarchy dissolves the connection between complexity and size.

These three statements about hierarchy are, of course, only the grossest sorts of generalizations. They would have to be modified in detail before they could be applied to specific organizational situations. They do provide, however, strong reasons for believing that almost any system of great complexity will have the rooms-within-rooms structure that we observe in actual human organizations. The reasons for hierarchy go far beyond the need for unity of command or other considerations relating to authority.

hierarchy in computing systems

None of these arguments for hierarchy is refuted by the characteristics of the new computer technology—either its software or its hardware. Consider first the software, the computer programs. Whenever complex programs have been written—whether for scientific computing, business data processing, or heuristic problem solving—they have always turned out to have a clear-cut hierarchic structure. The overall program is subdivided into subprograms; the subprograms are further subdivided; and so on. Moreover, in some general sense, the higher level programs control or govern the behavior of the lower level programs, so that we find among these programs relations of authority that are not dissimilar from those we are familiar with in human organizations.

Computer hardware shows equally clear evidences of hierarchy, from the level of individual circuits to the level of major components like memories, output devices, and central processors. Current developments in hardware, especially the movement toward building systems as assemblages of minicomputers, make hierarchic arrangements even more prominent. If a large system is to be built out of numerous component computing elements, then each element cannot be at all times in direct communication with each other element. Nor can detailed information about each component be distributed everywhere in the system. Hence, the larger our computing systems become, the more they are going to look, in their general structural arrangements, like human organizations. If they are organized hierarchically, their total complexity will not place increasing demands upon the complexity of the individual components.

Since organizations are designed to enable humans and their machines to accomplish goals, organizational form must be a joint function of human characteristics and the nature of the task environment. It must reflect the capabilities and limitations of the people and tools that are to carry out the tasks. It must reflect the resistance and ductility of the materials to which the people and tools apply themselves. What the preceding paragraphs assert is that one of the near-universal aspects of organizational form, hierarchy, reflects no very specific properties of human or computer, but a very general property. An organization will tend to assume hierarchic form whenever the task environment is complex relative to the problem-solving and communicating powers of the organization members and their tools. Hierarchy is the adaptive form for finite intelligence to assume in the face of complexity.

The conclusion I draw from this analysis is that the automation of decision making, irrespective of how far it goes and in what directions it proceeds, is unlikely to obliterate the basically hierarchic structure of organizations. The decision-making process will still call for departmentalization and subdepartmentalization of responsibilities. It is only through hierarchy that serial systems such as human beings can perform large tasks that require parallel activity on many fronts. Nevertheless, hierarchic structures can differ greatly among themselves with respect to the degree of centralization and decentralization of decision making, and with respect to the authority relations among their components. We turn to these matters next.

Centralization and Decentralization

One of the major contemporary issues in organization design is how centralized or decentralized the decision-making process will be—how much of the decision making should be done by the executives of the larger units, and how much should be delegated to lower levels. But centralizing and decentralizing are not genuine alternatives for organizing. The question is not *whether* we shall decentralize, but *how far* we shall decentralize. What we seek, again, is a golden mean: We want to find the proper level in the organization hierarchy—neither too high nor too low—for each important class of decisions.

Today, the terms "centralization" and "decentralization" are heavily laden with value. In general, decentralization is regarded as a good thing, and centralization a bad thing. Decentralization is commonly equated with autonomy, self-determination, or even self-actualization. Centralization is equated with bureaucracy (in the pejorative sense of that term) or with authoritarianism, and is often named as a prime force causing dehumanization of organizations and alienation of their members. In the last chapter we examined these kinds of claims closely enough to engender some skepticism about them.

trends toward decentralization

In the first twenty years after the Second World War there was a movement toward decentralization in large American business firms. This movement was probably a sound development, but it did not signify that more decentralization at all times and under all circumstances is a good thing. It signified that at a particular time in history, many American firms, which had experienced almost continuous long-term growth and diversification, discovered that they could operate more effectively if they brought together all the activities relating to individual products or groups of similar products and decentralized a great deal of decision making to the departments handling these products or product groups. At the very time this process was taking place there were many crosscurrents leading toward centralization in the same companies—centralization, for example, of industrial relations functions. There is no contradiction here. Different decisions need to be made in different organizational locations, and the best location for a class of decisions may change as circumstances change.

In the past ten years we have heard less about decentralization, and there probably has even been some net movement in the direction of centralization. Some of this reversal of trend was produced by second thoughts after the earlier enthusiasm for product-group divisionalization of company structures. Interdependencies among divisions were sometimes greater than had been supposed, and divisionalization did not always work out successfully. A second force working toward recentralization of decisions has been the introduction of computers and automation.

There are usually two pressures toward greater decentralization in a business organization. First, decentralization may help make the profit motive meaningful for a larger group of executives by allowing profit goals to be established for individual subdivisions of the company. Second, it may simplify the decision-making process by separating out groups of related activities —production, engineering, marketing, and finance for particular products—and allowing decisions to be taken on these matters within the relevant organizational subdivisions. Advantages can be realized in either of these ways only if the units to which decision is delegated are natural subdivisions—if, in fact, the actions taken in one of them do not affect in too much detail or too strongly what happens in the others. Hierarchy always implies some measure of decentralization. It always involves a balancing of the savings obtained through direct local action against the losses incurred through ignoring indirect consequences for the whole organization.

When the cable and the wireless were added to the world's techniques of communication, the organization of every nation's foreign office changed. The ambassador and minister, who had exercised broad, discretionary decision-making functions in the previous decentralized system, were now brought under much closer central control. The balance between the costs in time and money of communication with the center, and the advantages of coordination by the center had been radically altered.

The automation of important parts of business data-processing and decision-making activity, and the trend toward a much higher degree of structuring and programing of even the non-automated part, is also altering the balance of advantage between centralization and decentralization. The main issue is not economies of scale—not the question of whether a given data-processing job can better be done by one large computer at a central location or a number of smaller ones, geographically or departmentally decentralized. Rather, the main issue is how to take advantage of the greater analytic capacity, the larger ability to handle the interrelations of things, that the new developments in decision making provide. A second issue is how to deal with the technological fact that the processing of information within a coordinated computing system is orders of magnitude faster than the input-output rates at which one such system can communicate with another, particularly where human links are required.

centralization of interdependent decisions

Let us consider the first issue: the capacity of the decision-making system to handle intricate interrelations in a complex system. In many factories in the past, the extent to which the schedules of one department were coordinated in detail with the schedules of a second department, consuming, say, part of the output of the first, was limited by the computational complexity of the scheduling problem. Often the best we could do was to set up a reasonable scheduling scheme for each department and put a sizable buffer inventory of semifinished product between them to prevent fluctuations in the operation of the first from interfering with the operation of the second. We accepted the cost of holding the inventory to avoid the cost of taking account of detailed scheduling interactions.

We paid large inventory costs, also, to permit factory and sales managements to make decisions in semi-independence of each other. The factory often stocked finished products so that it could deliver on demand to sales warehouses; the warehouses stocked the same products so that the factory would have time to manufacture a new batch after an order was placed. Often, too, manufacturing and sales departments made their decisions on the basis of independent forecasts of orders.

These practices still prevail in some companies, but many have achieved substantial reductions in inventory costs by changing their decision-making systems. With the development of operations research techniques for determining optimal production rates and inventory levels, and with the development of the technical means to keep and adjust the data that are required to apply the optimizing procedures, large savings have been attained through inventory reductions and the smoothing of production operations, but at the cost of centralizing to a greater extent than previously the factory-scheduling and warehouse-ordering decisions. Since the source of the savings is the coordination of the decisions, centralization is unavoidable if the savings are to be secured.

The mismatch—unlikely to be removed in the near future—between the kinds of records that humans produce readily and read readily, and the kinds that automatic devices produce and read readily, is a second technological factor pushing in the direction of centralization. Since processing steps in an automated data-processing system are executed in a thousandth or even millionth of a second, the whole system must be organized on a flow basis with infrequent intervention from outside. Intervention more and more takes the form of designing the system itself —programing—and less and less the form of participating in its minute-by-minute operation. Moreover, the parts of the system must mesh. Hence, the design of decision-making and data-processing systems tends to be a relatively centralized function. It is a little like ship design. There is no use in one group of experts producing the design for the hull, another the design for the power plant, a third the plans for the passenger quarters, and so on, unless great pains are taken at each step to see that all these parts will fit into a seaworthy ship.

It may be objected that the question of motivation has

been overlooked in this whole discussion. If decision making is centralized, how can the middle-level executive be induced to work hard and effectively? First, we should observe that the principle of decentralized profit-and-loss accounting has never been carried much below the level of product-group departments and cannot, in fact, be applied successfully to fragmented segments of highly interdependent activities. Second, we may question whether the conditions under which middle management has in the past exercised its decision-making prerogatives were actually good conditions from a motivational standpoint.

Most existing decentralized organization structures have at least three weaknesses in motivating middle-management executives effectively. First, they encourage the formation of and loyalty to subgoals that are only partly parallel with the goals of the organization. Second, they require so much nonprogramed problem solving in a setting of confusion that they do not provide the satisfactions which, we argued in the last chapter, are valued by the true professional. Third, they realize none of the advantages that, by hindsight, we find we have often gained in factory automation from substituting system-paced for man-paced operation.

The question of motivation just raised has a relevance beyond the issue of decentralization. I will discuss it again later, in the section on authority and responsibility. Meanwhile, we can summarize the present discussion by saying that the new developments in decision making will tend to induce somewhat more centralization in decision-making activities at middle-management levels.

strategic planning and centralization

The spread in the use of complex models as an aid to decision making has not been limited to middle-management decisions. Increasingly, corporate planning at top management levels is being informed and assisted by a variety of computerized analytic techniques, including modeling of the firm itself and of its economic environment. Later, I will have something to say about these techniques, but for the moment I wish only to consider their implications for centralization. The growing use of modeling as a component of strategy formulation and strategic planning has led to some expansion of corporate planning staffs to carry out this function, producing new flows of information and advice from such staffs to top management. It is difficult to interpret this development as a traditional increase in centralization. What it principally means is that a considerable amount of managerial and technical effort all up and down the line, which previously was devoted to day-to-day decision making, is now devoted to the design of the decision process itself, and to developing and maintaining the basic models and data bases required for strategic analysis.

At all organizational levels, as decision processes become more explicit, and as their components are more and more embedded in computer programs, decisions and the analyses that underlie them become more and more transportable. If the method of analysis is explicit, and the informational and other premises that enter into it can be specified, then it does not matter very much at what organizational locations the analysis is carried out. It becomes increasingly feasible to carry out alternative analyses, using different assumptions and even different decision frameworks and analytic techniques, and employ them all as inputs to the final decision process. Since information, goal premises and constraints derived from all sorts of organizational and extraorganizational sources can provide inputs to the analytic processes, the locus of decision making becomes even more diffuse than it has been in the past. The organizational hierarchy remains as a critical mechanism for monitoring the process, but an increasing part of the flow of decisional premises runs across the boundaries of the formal hierarchy. Centralization is increased in the sense that decisions that have important interdependencies are less often made in independence of each other. But at the same time there is very broad participation in the making of decisions, through providing the information and premises on which they are based.

Authority and Responsibility

Since I use the term *authority*, in this book and elsewhere, in a wider sense than is common, I will say once more what I mean by it. A person accepts authority whenever he takes decision premises from others as inputs to his own decisions. Rewards and punishments provide the most obvious motives for accepting authority—especially, in organizations, the economic rewards associated with employment. But these are not the only motivating forces. Provided that a person is basically motivated to work toward the goals of an organization, much of the authority he accepts derives from the "logic of the situation." Decision premises are likely to be accepted if there is reason to believe that they are appropriate to the task and the situation. Expert advice is authoritative if it reflects the requirements of the situation. And a communication is frequently accepted as authoritative because it comes from an organizational source that is in a position to be "expert" for that kind of communication.

Closely related to the expertness of a source of decision premises is its legitimacy. The division of labor in an organization establishes expectations that certain kinds of decision premises will emanate from certain departments in the organization. A regulation about personnel practices has prima facie legitimacy if it comes from the personnel department.

Under some circumstances people chafe at accepting authority; under other circumstances they do not feel it as being in the least demeaning. In particular, authority is accepted more readily if it appears consistent with the logic of the situation than if it appears arbitrary or capricious. The experience of freedom and responsibility does not require complete independence from outside influence. Rather, it requires that the outside constraints and demands be understandable and reasonable. One does not feel unfree handling a sailboat, even though most of one's responses are governed by the moment-to-moment demands of wind and wave.

As the sailboat example illustrates, the physical environment is often as important a source of decision premises as are other human beings. One way to control a driver's behavior is to pass and enforce a speed law; another is to attach a governor to his motor or reduce its horsepower. Human reactions to authority are not particularly different as the authority resides in a human or in a physical source. Human beings react negatively to human authority that they view as inimical or frustrating to their goals; they also react negatively to rain at a picnic.

automation and authority relations

Many of the effects of automation upon the jobs of managers can be described in terms of the changes that are occurring in authority relations. Let me draw a sketch of the factory manager's job under the preautomated technology. How far it is a caricature, and how far a reasonably accurate portrait, I shall let you decide. What is the factory manager's authority? He can hire and fire (within the limits of union negotiations and company regulations). He can determine what shall be produced in his factory and how much (subject to company plans and schedules). He can make minor improvements in equipment and recommend major ones. In doing these things, he is subject to all kinds of constraints and evaluations imposed by the rest of the organization. Moreover, the connection between what he decides and what actually happens in the factory is often highly tenuous. He proposes, and a complex administrative system disposes.

For what is the factory manager held accountable? He must keep his costs within the budget standards. He must not run out of items that are ordered. If he does, he must produce them in great haste. He must keep his inventories down. His men must not have accidents. And so on.

Subject to this whole range of conflicting pressures, controlling a complex system whose response to instructions is often erratic and unpredictable, the environment of the typical middle-management executive—of which the factory manager is just one example—is not the kind of environment a psychologist would design to produce high motivation, or to reduce tension and frustration.

The manager responds in understandable ways. He transmits to his subordinates the pressures imposed by his superiors—he becomes a work pusher, seeking to motivate by creating for his subordinates the same environment of pressure and constraint that he experiences. He and his subordinates become expediters, dealing with the pressure that is felt at the moment by getting out a particular order, fixing a particular disabled machine, following up a particular tardy supplier.

Activities of pace setting, work pushing and expediting will not be absent from the highly automated factory or office. However, as automation and rationalization of the decision-making process go forward, these aspects of the managerial job are likely to recede in importance. The manager's work will become more professionalized. Both the authority he exercises and the authority exercised over him will seem to stem, more often than before, from the logic of the situation.

There are several reasons for predicting changes of this kind. First, the instrumentation of the automated factory or office, an integral component of the automated devices themselves, makes the behavior of the system more predictable and controllable. Second, there is less capriciousness in the application of policies and regulations. The factory manager is less likely to receive a sudden order to reduce inventories because the treasurer has awakened to a crisis in cash balances. Instead, both inventories and cash balances will be controlled by decision rules that recognize their interdependence and anticipate their fluctuations.

More generally, in the automated system, hourly daily decisions will require less and less human intervention. The main responsibilities of managers will be for the maintenance and improvement of the decision *system*, and the motivation and training of their subordinates. Line management, as we have known it in the past, will be a smaller part of the job, and managers will spend much of their time and effort as members of task groups engaged in analyzing and designing policies and the systems for implementing them.

the future of middle management

When computers were first applied to middle-management decisions, there were some predictions of the "withering away" of middle management. Organization charts, it was argued, would no longer look like pyramids, but like hourglasses, with many operative employees, and an increased number of higher-level managers, but very little in between.[4] This did not happen—for

[4] See Harold J. Leavitt and Thomas L. Whisler, "Management in the 1980's," *Harvard Business Review*, 36:41-48 (November-December 1958).

several reasons. First, automation decreased the number of operative employees substantially. Second, although the demand for line management was substantially smaller, middle managers were needed for the new staff operations of designing and maintaining the automated decision-making and planning systems. Staff units grew at the expense of line units.

The net result of these developments has generally been to reduce the number of layers in organizations as line organizations became smaller, but to broaden organizations by expanding the staff functions. The resulting structures are in some respects more complex than those that preceded them: There is more interaction among units across hierarchic boundaries. But this complexity is tolerable because it is less concerned with high-frequency interaction required for the real-time operation of the system, and more concerned with planning activities having longer time horizons.

The changes in management and organization structure described in the preceding paragraphs are no longer just matters of speculation. Although the movement toward automization is still in an early stage of development, we now begin to have some case studies and a few systematic investigations to support the predictions. The book by Thomas L. Whisler [5] has provided a thoughtful overview of the evidence. The data reveal a shrinkage of line organizations in companies that have introduced computers, some reductions in the span of control of executives, some recentralization of interdependent decisions, and a reduction in the number of levels of authority. A few data show the growth in staff responsibilities and the shrinkage of line responsibilities of managers. Although the evidence on all these matters is not voluminous, it is fairly consistent internally, and consistent also with the predictions made from the characteristics of the technology.

Information Systems and Planning

When computers are introduced into organizations, one of the things they are supposed to provide is more and better in-

[5] Whisler, *Information Technology*, especially Chapters 4 and 5.

formation. One way that is often proposed for reaching this goal is to construct a management information system. From our earlier discussion of the nature of an information-rich world, we should be skeptical of proposals to provide us with more information, although we might well want better information. In this section we consider the merits of management information systems and other techniques for bringing information to bear on top management decisions.

the uses of information

Every company takes considerable pains to ascertain its profit periodically. What questions is this information supposed to answer? First, it tells the company how well it is doing—it serves as a crude score card of corporate well-being.

Second, the profit figure is used to call attention to problems when they arise. When profit drops from last year's level, a company looks to needed shifts in its policies.

Third, the profit, along with other statistics, may become grist for the mill of research that is trying to understand the structure of the company or its industry—how advertising expenditures affect sales, for example. This use does not aim at immediate action, but at understanding system response, to make subsequent actions more intelligent.

Fourth, the profit, together with other measures, may be used to ascertain the present state of the system in order to plan for its future—to predict, for example, future sales and profit levels in order to determine the need for capital expenditures to increase capacity.

Information has, at least potentially, these four broad classes of uses: as score card, to direct attention, to analyze a system's structure and dynamics, and to ascertain parameters of its current state. A single statistic or set of statistics may serve, at different times, in all four uses. On the other hand, data of particular kinds may be especially appropriate to one of these uses.

An information system designed to provide a set of score cards or of attention directors will not necessarily be well suited to gaining a deeper understanding of system dynamics; the converse is undoubtedly also true. Design of an information system must begin with specifying what questions the information is to answer, and for what levels of management. Those specifications, in turn, must be derived from an understanding of how and where decisions are made in the organization.

Let us see how well this has been done for management information systems and for strategic planning systems, the two most common approaches to building information systems for management decision.

management information systems [6]

There is no single, definite description of just what a management information system is. The phrase has to be understood sociologically rather than technically. In the early sixties, as computers were introduced into more and more companies for accounting purposes (i.e., mainly for score-carding and attention-directing uses), an interest arose in putting the information produced by those systems to other management uses. The management information system, as typically conceived, was a byproduct of the availability of information gathered initially for other purposes. The design of management information systems most often began with asking what could be done with the information that was already there, not with asking what decisions were being made, and what information would be helpful in making them.

Early management information systems produced and distributed voluminous reports filled with information that management previously could not easily get at. That the reports were not used very much occasioned more surprise among their producers than among the intended users. The designers of the system had not learned the first lesson of living in an information-rich world: that a major task of an effective information system is to filter information, not to proliferate it.

[6] For a survey of the scanty empirical literature on this topic, see R. L. Van Horn, "Empirical Studies of Management Information Systems," *Data Base*, 5:172-82 (Winter 1973); and H. C. Lucas, Jr., *Why Information Systems Fail* (New York: Columbia University Press, 1975). The views expressed in this section are largely based on my own knowledge of computer technology and information systems, and on conversations with executives in companies that have introduced such systems. I think I am reporting a consensus from which there is little dissent.

The garrulousness of these management information systems was fairly easily remedied by selective distribution of reports, and by rediscovery of the principle of exceptions: Only departures from the norm, the unexpected, need to be communicated. Of course, in order to use the principle of exceptions, normal ranges for statistics had to be stored in the computer so that it could detect automatically when a statistic fell outside the expected range and should be reported. Adding such norms to an information system was not a difficult technical problem. When the paper flow produced by management information systems had been stanched, a new problem appeared. The reports produced by such systems were frequently found useful by some executives in middle-management positions, but were of much less interest to top managers.

The result has been that much of the original enthusiasm for management information systems has evaporated, and a number of systems already developed have been abandoned. This outcome could have been predicted (and in some cases was) by starting the analysis where every analysis of an information system should start: with the jobs of the persons the system is supposed to serve. If we analyze the decisions in which top managers play a major role, and the kinds of information that would be most relevant to making those decisions well, we soon see that a large part of the information that is wanted does not originate in company records, or inside the company at all. Top managers stand at the interface between a company and its environment of customers, competitors, sources of funds, and political and social systems. Their main concerns are to discover how the company can live effectively and profitably in this external environment. A large part, probably most, of the information that would be genuinely helpful to them is information about that external environment, obtainable only from external sources.

One kind of external information of importance to top management is to be found in news magazines, trade journals, financial journals, journals of economics and politics, scientific journals, and books. A second kind of external information is to be found in governmental and private statistical compilations dealing with the international or national economy or with particular industries. Corresponding approximately to these two kinds of information sources are two kinds of uses of information, which we may call *intelligence* and *strategic planning,* respectively. I will say a few words about intelligence in this section, and about strategic planning in the next.

Intelligence information (obviously, I am not restricting this term to information obtained by clandestine means) is mainly used for attention-directing and parameter-measuring purposes. It helps management determine where they are and what problems need attention—particularly problems originating from changes in the external environment. The design of modern information systems for intelligence purposes is in its infancy. There are at least two reasons why the development has been so slow. The first has to do with the gathering of the information, the second with its selective dissemination.

Most of the information one would want to gather into an intelligence system is available only as man-readable, natural language text. Converting it into machine-readable form so that it can be ingested by a computer is a difficult technical problem. Optical scanning devices are available today that can accept text within a range of type faces, but the process is, at best, costly.

This problem of access to the information sources is severe, but temporary. We have already observed that there is no technical reason why any material that is produced in printed, man-readable form should not be produced simultaneously in machine-readable form, on magnetic tape or otherwise. Virtually everything that is printed today passes at some time through a mechanical system—a linotype or monotype machine, an automatic typesetting machine, or a typewriter. At that moment, machine-readable text can be prepared as a byproduct, at almost no cost. As soon as there is an economic demand for the machine-readable version of printed material, it will be produced. Already, there is a growing supply of such material—company financial information, for example, legal citations, bibliographical references and abstracts, and others. The existing Associated Press system for news transmission is an excellent example of the new technology.

Because of the large initial startup costs, and the costs of debugging a new technology, this transfer has been very slow in getting under way—all the elements of the technology have been available for ten or twenty years. But we may expect a snowball-ing effect. The more such material becomes available, the more potential users will find it advantageous to equip themselves to handle it. And the more potential customers who are equipped to handle it, the more profitable it will become to produce it. The snowball has gained considerable velocity in the past two or three years, and we may expect it to gain greatly in volume—and speed—in the next five.

The second technical problem that must be solved in an automated intelligence system is to filter the information and dis-tribute it selectively. Not all of the filtering need be done in-ternally, within the company. The thousands and thousands of specialized trade and technical journals and newsletters that are published are specialized information filters, designed for their own particular audiences. They too will increasingly use the new technology to carry out their task more efficiently.

At present, the capabilities of computers for filtering natural-language text are also relatively primitive. For most purposes, however, filtering can be done by a system that falls far short of "understanding" the material it is scanning. Techniques for iden-tifying key words can in many instances select out relevant infor-mation as effectively as can be done by a human reader scanning large volumes of text rapidly. Hence, there are no deep technical reasons why computers should not augment human capabilities in performing intelligence functions.

Several important sources of information to top managers—correspondence and conversation—still lie beyond the reach of automatic systems. The computer has no present way of joining in the conversation on the tee or in the locker room. The ques-tion before us, however, is not whether intelligence systems can be automated completely, but whether there is an important role in such systems for the new information technology. It is clear that there is.

strategic planning

Concern with strategic planning, unlike concern with manage-ment information systems, began with decisions looking for in-formation rather than information looking for decisions. Hence, efforts at strategic planning have avoided some of the mistakes that were made in designing management information systems (and have perhaps made some others of their own). To an important extent, information-gathering for strategic planning has taken the form of constructing computer models to simulate either the business firm, or some part of its economic or social environment. The General Electric Company, for example, beginning in the late fifties, has undertaken to examine and model social trends relevant to the demand for its products. Many companies today make some use of economic forecasting models, either produced internally or obtained from external consulting sources.

While intelligence systems are generally designed for attention-directing and parameter-estimating purposes, strategic planning is more concerned with gaining an understanding of system behavior. Losing sight of this objective, and of the decisions that are to be made, sometimes leads strategic planning efforts into an excessive preoccupation with prediction.

Prediction for its own sake is a costly and pointless game. The objective in making predictions and projections into the future should be to provide a basis for the decisions that are to be taken today; tomorrow's decisions can and should be made on the basis of the information available tomorrow. Some decisions are required today to provide lead time for tomorrow's actions.

The future is relevant to decisions taken today only to the extent that these decisions have consequences for the future that are in some sense irreversible—that cannot be undone. The main reason, for example, why forward planning is so closely associated with decisions about physical structures is the permanence of those structures, and the long future period during which the design decisions have consequences that cannot be altered without cost. The higher the interest rates we use in discounting remote consequences, the less sensitive are our decisions to predictions of the future, hence the less important it is to make such predictions.

Since the development and diffusion of new innovations takes considerable time—years and decades—there is usually little need for "blue-sky" prediction in a decision system. Using an aggregate model of a system to predict from advances in science that have already occurred to their technical, economic, and social consequences can often be important means for improving the accuracy of those predictions that are really crucial to planning. (That is what this book is all about, as applied to advances in computers and the theory of decision making.) Strategic planning directed toward key factors in decision makes quite different information demands from planning conceived as "comprehensive" prediction and control of a system. For the former, the greatest need is for thorough understanding of system structure rather than for masses of detailed data on the current state of the system.

summary: information systems

The initial efforts toward designing management information systems started with available data rather than with decisions to be made. In general, these systems, even as revised with experience, have been disappointing, and more relevant to the jobs of managers at relatively low levels in the organization than to the jobs of top managers. A major reason for this outcome is that higher management must turn its gaze mainly outside the organization, and is less concerned with data that originate within its boundaries than with information from the environment. The information systems that are likely to be of most importance for top management are those that gather and filter intelligence from external sources, and those that are designed to assist strategic planning efforts. These are the real "management information systems," though they seldom bear that label.

The technology is now available for constructing significant automated systems to handle a variety of information expressed in natural language and available from books, magazines and newspapers. What is chiefly slowing the introduction of such systems is delay by the producers of these sorts of information in making it available in computer-readable form. That situation is now changing.

Computers have been used in a variety of ways, in conjunction with planning activities, to model company and industry operations and the economic environment. A common, but remediable, weakness in many uses of computers in strategic planning is an overemphasis upon detailed forecasting as a component of the planning process. While some forecasting is an essential element of most planning, the effective use of planning systems calls for ingenuity in avoiding the dependence of plans upon projections at a level of detail that is not attainable. Computers, no matter how powerful, do not permit accurate forecasts to be made of the future of systems whose underlying structure and dynamics are not fully understood—a stricture that certainly applies to the economic system and its major components.

A Final Sketch of the New Organization

We have now surveyed the changes that are taking place in the manager's job as the organization takes on more and more of the aspects of a complex man-machine system. We have looked at the kinds of information systems that are likely to aid him in his work. If a couple of terms are desired to summarize the direction of change we may expect in management, I would propose "rationalization" and "professionalization." The new information technology has already produced large changes in the decision-making processes of middle management in these directions. With the development of more sophisticated systems for gathering and filtering external information and for modeling strategic plans, these effects are going to be felt increasingly at the higher levels of management.

Basic research, like that described in Chapter 2, into human thinking and nonprogramed decision making is still far in advance of anything that has been applied to the practice of management. For this reason, and because the research continues to move forward rapidly, we should expect the long-range transformation of top-management practice to go far beyond the external intelligence systems and strategic planning systems that can be foreseen for the relatively near future. This chapter has focused mainly on the foreseeable middle run rather than the long-run prospect, whose outlines are still very dim.

In terms of subjective feel, the changes now taking place mean that the manager will find himself dealing more than in the past with a well-structured system whose problems have to be diagnosed and corrected objectively and analytically, and less with unpredictable and sometimes recalcitrant people who have to be persuaded, prodded, rewarded, and cajoled. For some managers, important satisfactions derived in the past from certain kinds of interpersonal relations with others will be lost. For other managers, important satisfactions from a feeling of the adequacy of their professional skills will be gained.

My guess, and it is only a guess, is that the gains in satisfaction from the change will overbalance the losses. I have two reasons for making this guess: first, because this seems to be the general experience in factory automation, as that affects supervisors and managers; second, because the kinds of interpersonal relations called for in the new environment seem to me generally less frustrating and more wholesome than many of those we encounter in present-day supervisory relations. Man does not generally work well with his fellow man in relations saturated with arbitrary authority and dependence, with control and subordination, even though these have been the predominant human relations in such settings in the past. He works much better when he is teamed with his fellow man in coping with an objective, understandable, external environment. That will be more and more his situation as the new techniques of decision making come into wide use.

When we assemble all the evidence about the new information technology and the applications it has already had in business organizations, we see that the facts do not support the anxieties that are so often and so unnecessarily aroused by the stereotype of the robot. These anxieties are unnecessary because the existence in the world today of machines that think, and of theories that explain the processes of human thinking, subtracts not an inch, not a hair, from the stature of man. Man is always vulnerable when he rests his case for his worth and dignity on how he differs from the rest of the world, or on the special place he has in God's scheme or nature's. Man must rest his case on what he is. This is in no way changed when electronic systems can duplicate some of his functions or when some of the mystery of his processes of thought is taken away.

The changes I am predicting for the decision-making processes in organizations do not mean that workers and executives

will find the organizations they will work in strange and unfamiliar. On the contrary, in many of their important aspects the
new organizations will much resemble those we know now.

1.   Organizations will still be constructed in three layers: an
underlying system of physical production and distribution processes, a layer of programed (and probably largely automated)
decision processes for governing the routine day-to-day operation
of the physical system, and a layer of non-programed decision
processes (carried out in a man-machine system) for monitoring
the first-level processes, redesigning them, and changing parameter values.

2.   Organizations will still be hierarchic in form. The organization will be divided into major subparts, each of these into
parts, and so on, in familiar forms of departmentalization. The
exact bases for drawing departmental lines may change somewhat.
Product divisions may become even more important than they are
today, while the sharp lines of demarcation among purchasing,
manufacturing, engineering, and sales are likely to fade.

But there is a more fundamental way in which the organizations of the future will appear to those in them very much like the
organizations of today. Man is a problem-solving, skill-using, social
animal. Once he has satisfied his hunger, two main kinds of
experiences are significant to him. One of his deepest needs is to
apply his skills, whatever they may be, to challenging tasks—to
feel the exhilaration of the well-struck ball or the well-solved
problem. The other need is to find meaningful and warm relations with a few other human beings—to love and be loved, to
share experiences, to respect and be respected, to work in common tasks.

Particular characteristics of the physical environment and
the task environment are significant to man mainly as they affect
these needs. The scientist satisfies them in one environment, the
artist in another; but they are the same needs. A good business
novel or business biography is not about business. It is about love,
hate, pride, craftsmanship, jealousy, comradeship, ambition, pleasure. These have been, and will continue to be, man's central
concerns.

The automation and rationalization of decision making will, to be sure, alter the climate of organizations in ways important to these human concerns. I have indicated what some of the changes may be. On balance, they seem to me changes that will make it easier rather than harder for the executive's daily work to be a significant and satisfying part of his life.

5

economic impact
of
the information technology [1]

The scope of this final chapter is a little broader than the word "economic" in its title may suggest. In the past, the principal concern associated with new technologies has been the possibility that they might cause widespread unemployment. Today, there is an even greater concern that they may cause the exhaustion of vital resources or of the Earth's capacity to absorb the polluting byproducts of production. Both of these issues will be discussed here, with particular attention to the specific characteristics of information processing technology and the ways in which it differs from other technologies. The first part of the chapter discusses the issue of overproduction and unemployment, the second part,

[1] I am deeply indebted to a number of friends, including G. L. Bach, M. Bronfenbrenner, M. Joseph, M. Kamien, L. Lave, and L. Rapping, for numerous enlightening discussions on the subject of this chapter. Needless to say, they are not responsible for my conclusions.

the problems of resource exhaustion and pollution, and the special role the information processing technology may have in dealing with those problems.

One of the interesting minor features of the depression that reached its climax in 1974–75 is that almost no one blamed it, and the attendant unemployment, upon automation or computers. In earlier periods of unemployment, including the Great Depression, it was often charged that technology was the villain. Perhaps in the present case there were too many other candidates for that role: inflation, the energy crisis, the Administration's tight-money policy, the contagion of economic difficulties abroad. On this occasion, antitechnological attitudes have expressed themselves in concern about resources and the environment instead of concern about employment.

Nevertheless, the belief is still often expressed in books and articles on computers and automation that introduction of these technologies causes unemployment. That belief has now been joined by the belief that the continuation of the technological advance we have had during the past two centuries is barred by resource and environmental limitations. The first of these beliefs makes no distinction among particular technologies—what is at issue is simply the economic impact of increasing productivity, whatever the cause of that increase. An analysis of resource and environmental limitations, however, depends very much on the specific kind of technological change we are considering. The new information technology has very different implications for resources and environment than the technologies that are based on the application of energy. For this reason, the discussion of technological change in the first part of the chapter will be quite general; while in the second part, it will refer specifically to computers and automation, and their application to managerial and social decision making.

Technology and Unemployment

For over a century in Europe, America, and Japan, output per worker and amount of capital per worker have been increasing steadily. Throughout most of this period employment of about 95 percent of the labor force has been the rule, and periods of mass unemployment the exception. There has been no discernible long-term trend in percentage of unemployment. Real wages of workers have risen steadily and substantially. These are the central facts of the Industrial Revolution.

The costs that this Revolution sometimes imposed on particular groups in society, including labor, have been often acknowledged and often described. Admitting these costs, it is generally agreed that the Industrial Revolution created for the first time a realistic expectation that abject poverty might be banished from the world. There remained strong disagreement about what institutions would have to be established to achieve the increased productivity and to share its benefits widely. Apart from orthodox Marxists, however, most analysts who studied this history agreed that the Industrial Revolution had, in the long run, increased the real income of most segments of the population.

Yet optimism has not been the universal tone of commentary on the Industrial Revolution. At any given moment in history, most observers grant its past benefits without necessarily being willing to project these benefits into the future. Thus, while advances in technology and the accumulation of capital are credited with past improvements in wages, continuing advances in technology and capital accumulation predicted for the future cause grave doubts and anxieties. These doubts and anxieties center on three possibilities: that productivity may reach such high levels that the goods and services produced can or will no longer be consumed; that production will place insupportable demands upon resources or upon the pollution-absorbing capacity of the environment; and that machines may become so efficient that labor will face massive unemployment. The last worry is probably the earliest of the three to arise, for the idea of a glut of goods or of resource exhaustion could hardly have occurred to anyone who had not experienced twentieth-century America. The concern over technological unemployment was expressed vigorously more than a century ago by the Luddites in their attempt, by force, to prevent machine weaving.

No one would question the possibility of overproduction and underemployment in the short run. Since we have experienced it, it would be fruitless for our theories to deny it, and, since the time of Keynes, they no longer do. No one would doubt, either, that short-run overproduction and underemployment may cause harm to large numbers of people, and that it is one of the central objectives of governmental economic policy to prevent this harm. It is a sign of significant social progress that our generation is less willing than any previous one to impose the costs of social and technological change on those who happen to be displaced or disadvantaged by it.

But the expressions of worry and concern about technological change are not limited only to the short-run dislocations it produces. It is often asserted or implied that continuing increases in productivity and mechanization may create permanent overproduction and underemployment. Is this a real possibility?

some facts of the matter

Reasoning is only as good as the premises on which it stands. Before we embark on our analysis of the economic effects of automation, it will be useful to establish two basic facts about the matter. It is particularly important to do so because much of the writing on automation has ignored easily available facts in favor of plausible but incorrect assumptions.

1. At the current rate of technological change in the United States, real net national product per capita is increasing at perhaps 2 or 3 percent per year. About half of this increase can be attributed to the continuing increase in capital and the other half to technological advance. This rate is only slightly higher—if, indeed, it is higher at all—than the rate sustained by the economy through the first half of this century. The widely held notion that productivity, driven by automation, is rising at breakneck speed is simply false.[2]

The meaning of a 3 percent annual increase in per capita productivity can be expressed in several ways. It means that 3 percent of the labor force (about 2 million workers) could be

---

[2] For a succinct summary of the facts on productivity, and references to the literature, see John T. Dunlop, ed., *Automation and Technological Change* (Englewood Cliffs, N.J.: Prentice-Hall, 1962), Chapter 7; and the excellent statistical analysis by John W. Kendrick and Ryuzo Sato, "Factor Prices, Productivity, and Economic Growth," *American Economic Review*, 53:974-1003 (December 1963). The latter work has now been brought up to date in J. W. Kendrick, *Postwar Productivity Trends in the United States, 1948-1969* (New York: National Bureau of Economic Research, 1973). See especially Chapter 4, "National Productivity and Economic Growth."

dispensed with each year (through increased leisure or unemployment) while holding average living standards constant. But it also means, of course, that if living standards are to rise at even a modest rate, this labor will be needed. While new workers entering the labor force swell the ranks of workers, they and their families also swell the ranks of consumers, hence do not alter the implications of the productivity figures—per capita incomes will rise only if these workers as well as the present labor force are employed.

2. Empirical case studies of automation do not reveal any general tendency toward either the upgrading or the downgrading of job skill requirements.[3] Hence, there is no empirical basis for claims that "a major problem associated with current technological change is the imbalance between the type of labor force our new technology increasingly requires and the skills and qualifications of the present labor force and of new entrants."[4]

To be sure, a larger and larger percentage of jobs are being filled with persons holding high school, college, and graduate degrees. There are more such people available, and, other things being equal, persons with more education are preferred to persons with less education by personnel recruiters—with scanty or no evidence that the educational requirements serve as more than a generalized test of intelligence and diligence. There is no evidence that a highly automated economy could not be operated efficiently with an educational profile like that of the western European or Japanese work forces—that is, with far less formal education than now prevails in the United States. Today, in fact, the opposite concern has arisen in this country: that the economy will not absorb the vast numbers who are being graduated from college. This is an issue we will address later in this chapter.

a glut of goods and leisure?

In the United States at the present time, the median family income is in the neighborhood of $12,000 per year. No one seriously

---

[3] We have already discussed this point in Chapter 2.
[4] Quoted from the Final Report of the 21st American Assembly, in Dunlop, *Automation,* p. 177.

questions the ability of families to spend incomes of that magnitude. If that ability were questioned, the doubts could readily be resolved by observing that they are being spent.

We have seen that a reasonable estimate of the current rate of increase of productivity per capita, and hence income per capita, would be about 3 percent per year. A continuing increase at this rate would double annual income per capita in about twenty years, that is, in one generation. Could families in the American economy, twenty years hence, consume income at a median rate of $24,000 per family; forty years hence, at a median rate of $48,000 per family?

Enough facts are already available to answer these questions. There are today, in the American society, substantial numbers of families whose incomes exceed $24,000 per year and $48,000 per year. These families are not experiencing difficulties in disposing of their incomes. Their savings, as a percentage of income, may be somewhat higher than the savings of the lower-income families, but a large fraction, at least, of these savings are later consumed in the form of retirement income. However that may be, there is no more reason to suppose that there is a limit on the capital formation the economy is able to absorb than to suppose that there is a limit on incomes.

Concern, then, about "excessive" incomes can be justified only by some assumption that families that now have low incomes are not capable of adjusting their wants upward to match those of the families already able to satisfy these larger wants. Low-income families might want to challenge that assumption.

Sometimes concern is expressed not about a prospective glut of services but about a glut of leisure. As productivity increases, some of the additional income takes the form of increased leisure—a shorter work week. Most persons who are alarmed at this prospect do not find that they themselves are endowed with too much leisure. But there are "many people," it is argued, who would not know what to do with leisure time, and who, presumably, would lend their hands to the Devil.

Most facts indicate that working-class leisure was significantly diminished by the Industrial Revolution, and that an increase in leisure even beyond the amounts we now enjoy would be required just to return to the conditions that prevailed in preindustrial Europe, or in many unindustrialized societies today. Although this does not decide whether such a return would be desirable, there is no reason to think that modest increments in leisure can not be absorbed. Although we do not know how long it would take mankind to adjust—if it ever would—to a world in which human labor was unessential to the production of goods and services, our previous analysis of productivity increases does not suggest that such a world will soon exist.

All that has been said thus far has been addressed to conditions in America and the other developed nations. Of course if we look at the underdeveloped three-quarters of the world, the notion that productivity will soon outstrip human wants becomes ludicrous. In those other parts of the world, the question is whether we can stave off widespread starvation long enough to bring the world's population in balance with its ability to meet basic human needs. I have emphasized the American situation, rather than the quite different one that faces much of humanity, in order to deal in its most extreme form with concerns about satiety.

of resources and the environment

Within the past ten years, the worry that the flood of production might drown us has been somewhat displaced by the worry that our productivity will exhaust our and the world's resources, or— what amounts to the same thing—the capacity of the environment to absorb the polluting byproducts of our production and consumption. These concerns require, of course, the most serious attention—it is rather remarkable that our society managed to ignore them as long as it did. It is important that we state accurately just what they imply for technological progress in general and for mechanization and automation in particular. Let me try to summarize the situation in a few bald propositions, leaving a fuller discussion to the final section of this chapter.

1. It is probable that limits on resources (from now on I will use that term to include limits on the absorptive capacity of the environment) will force a stabilization of world population in the quite near future—either by intent or by disaster.

2. It seems quite probable, also, that resource limitations will place an upper bound, even for a stabilized population, on the amounts of physical material that can be embodied in goods or transformed and excreted to the environment.

3. To make a wild guess, that upper bound might imply physical production at no more than ten times the present world rate, or a per capita rate somewhat less than the current United States average. (If I have placed the bound too low by a factor of five, the general argument is not much changed, though the time scale is extended.)

It is an easy, but fallacious, jump from these premises to the conclusion that technological progress, including automation, must soon come to a grinding halt. The conclusion is fallacious because it implies that technological advance necessarily means producing greater quantities of physical structures and commodities. All that technological advance implies is that a given quantity of output will be produced with a smaller input of capital and labor. As the world reaches levels of production and consumption like those in the United States, we may expect that a larger and larger fraction of the growth will take the form of human services, rather than being encapsulated in physical goods. Indeed, this shift from goods to services has already been going on for some time in our economy. Blue-collar workers (including farmers and farm workers) now make up less than 40 percent of the American labor force, as compared with 80 percent or more a century ago.

As resource limitations begin to press more severely on the American and world economies, we will have to begin to make a distinction between advances in technology and advances in productivity. Considerable technological progress (e.g., new and non-polluting sources of energy) will be needed just to enable us to maintain present levels of production without resource exhaustion or environmental damage. As we now do the bookkeeping for our national accounts, costs that are external to firms and households tend to be ignored. Resources devoted to reducing these costs do not show up on the accounts as increases in productivity.

Finally, we must attend to an important difference between

advances in the technology of information and advances in the older technology of energy. The conservation laws of physics tell us that we cannot have larger outputs of mass or energy without larger inputs. There is no such conservation law for information. As the rapid trends of the past decade toward miniaturization of computer components have shown, advances in the computer technology imply smaller, and not larger, inputs of metal or energy for a given amount of information processing output.

In summary, the prospect of greater resource scarcity in the future does not imply that our human resources cannot and will not be fully employed in producing goods and services. It probably does imply a continuing shift from the production of goods to the production of services. To an important degree, our advances in technology will be used in the future to combat rising costs due to resource scarcity.

The Economics of Employment

Today, we hear the term "automation" more often than "mechanization." From an economic standpoint, the distinction is unimportant. Automation is simply a continuation of that trend toward the use of capital in production that has been a central characteristic of the whole Industrial Revolution. What is possibly new, if anything, is the extension of mechanization to wider and wider ranges of productive processes, and the growing prospect of complete mechanization—that is, the technical feasibility of automatically controlled production processes that do not require human participation for their operation.

The argument against mechanization has always been that it causes technological unemployment because men cannot compete against increasingly efficient machines. The standard rebuttal of economic theory to this worry is to invoke the principle of comparative advantage. This principle shows, essentially, that both people and machines can be fully employed regardless of their relative productivity. By adjustment in the relative price of labor and capital, respectively, it will come about that labor will be employed in those processes in which it is relatively more productive, capital in the processes in which *it* is relatively more productive. Absolute rates of productivity are irrelevant, and, indeed, because of the noncomparability of labor units with capital units in physical terms, absolute rates cannot even be compared in an operationally meaningful way.

Although the doctrine of comparative advantage is a perfectly sound piece of reasoning, which applies under very general circumstances,[5] it does not settle all the essential issues. In particular, although it shows that at some wage all labor would be employed in equilibrium, no matter how efficient machines become, it does not predict what that wage would be. It does not guarantee that real wages will not drop as the economy's productivity improves through mechanization. It does not even guarantee that real wages will remain above the subsistence level.

The possibility that real wages might drop below the subsistence level as a result of automation is given a certain plausibility and poignancy by the almost complete disappearance of the horse as a factor of production (except at race tracks, hunt clubs, and riding stables). In 1915, the horse population in the United States reached its peak of some 21 million animals. By 1960, it was down to 2 million, almost none of them work horses. In the face of the farm tractor, the horse simply was not able to produce enough to pay its keep. Can the same thing happen to the human worker as automation proceeds?

To pay his way while he plowed a field, Dobbin had to do enough plowing to cover the interest and wear and tear on the plow. He also had to pay the wages of the man who drove him. Only after these two costs had been deducted was the remaining product available to pay his own keep. The introduction of the tractor did not make Dobbin physically less productive. A man and a horse and a plow could still plow as many acres a day as they ever could before. What the tractor did was increase the cost of the driver, whose productivity was now greater when he plowed by tractor. At the new real wages the man could command because of the invention of the mechanized substitute, the horse could not afford to pay for his services. He could no longer afford to support his master at the new level of luxury to which mechanization had accustomed him.

[5] It does not, of course, deny the possibility of short-run unemployment, but our present concern is with long-term effects.

Can the same thing happen to man in his symbiosis with the machine? Will man find himself in a position where he will be unable to pay for the help of machines in his production and have enough left over to maintain his present real wage? The introduction of new methods of machine production does not reduce the physical productivity of the old methods, but the old methods can become uneconomical if relative prices change. Our next step, therefore, must be to see whether we can say anything about the consequences of mechanization or automation for the prices (in real terms) of capital and labor.

the price of capital

The price of capital is interest, which, of course, is usually measured as a rate per unit of time per dollar value of the capital. Prices of most other things are measured in terms of dollars per physical unit. We need to make a similar translation of the rate of interest.

First, how can we measure the quantity of capital? Dollars are an unsatisfactory unit for this purpose, since we wish to make comparisons between different points of time, and therefore wish to deal with real—i.e., physical—quantities. The task of definition is made more difficult by the fact that the quality of capital, its productivity, may change over time as a result of the very technological change we are studying. Somehow, we must replace these rubber yardsticks with invariant ones.

To make matters more concrete, we shall develop the definitions in the context of a particular, if highly simplified, hypothetical economy. Although this economy is very simple, nothing has been left out of it that would change the conclusions to which our reasoning will lead.

In our hypothetical economy, only a single commodity, which we shall call beanbricks, is produced. Beanbricks serve both as the sole consumption good (they are eaten), and as the sole form of capital (they are burned as fuel used in producing more beanbricks). The other productive resource in the economy, besides beanbricks, is labor. A number of processes exist for making beanbricks, using various amounts of labor and various amounts of beanbrick fuel. Some of these processes are labor intensive—they use relatively large numbers of man-hours of labor, and relatively small numbers of beanbricks, to produce each thousand beanbricks. Other processes are capital intensive, using relatively smaller numbers of labor hours, but larger numbers of beanbricks as fuel, for each thousand beanbricks produced. Which processes will be profitable, and hence actually used, will depend on the relative prices of beanbricks and labor.

Technological change in this economy is represented by a reduction, for one or more production processes, of the number of labor hours and/or the number of beanbricks required to produce an output of a thousand beanbricks. A technological improvement may even increase one of these inputs if the other one is reduced sufficiently so that the net cost of inputs is lowered (substitution, in production, of beanbricks for labor, or labor for beanbricks).

The economy as described still does not require, or provide for, a rate of interest. To introduce interest, we suppose that beanbricks have to be aged for a year before they are used for fuel. Producers borrow beanbricks for fuel under an agreement to return them at the end of a year with interest (1.05 beanbricks in return for each beanbrick borrowed, say). We can say, then, that the price of a beanbrick is $(1+r)$ beanbricks, where r is the annual rate of interest. If producers are willing to offer more than this for beanbricks, owners will supply them in larger numbers and eat fewer. If producers offer less for beanbricks, they will be eaten instead of loaned. Hence, the price of beanbricks can never depart very far from $(1+r)$ beanbricks, measured in real (i.e., beanbrick) units. What is usually called *improvement in quality* of capital simply means, in this model, that beanbricks are used in more efficient production processes, processes that burn fewer beanbricks per thousand beanbricks produced. In this model, improvement in the effectiveness of beanbricks as fuel cannot and need not be distinguished from improvement in the processes for burning them.

Finally, we assume that the quantity of beanbricks that can be produced by any process is limited only by the quantities of labor and beanbrick fuel that are employed in that process. We assume constant returns to scale—proportionality of output to input—which might be a dubious assumption in an agricultural economy with limited arable land, but is probably a quite realistic first approximation to the kind of industrial economy in which we live.

real wages in equilibrium

With these preliminaries out of the way, we can trace out the effects of improvements in technology upon real wages in our hypothetical economy. Readers who want only the conclusions, and are uninterested in the methods of arriving at them, can skip to the section titled "Restatement."

For any production process, or average of processes, employed at equilibrium, the value of the output will equal the sum of the costs of the labor and capital inputs. That is to say, the real wage rate (measured in beanbricks) multiplied by the labor input requirement per beanbrick produced, plus the capital input requirement per beanbrick, equals the real price of a beanbrick, which, by definition, is unity:

(Equation 1) $\qquad p_L a_1 + (1 + r) a_2 = 1,$

where p_L is the wage rate, a_1 the labor input coefficient, r the rate of interest, a_2 the capital input coefficient. We assume that the total initial quantity of labor is L, the quantity of capital C, and the total output P. If a_1 and a_2 are average values for the economy ($a_1 = L/P$ and $a_2 = C/P$), the total value of production (the number of beanbricks produced) will simply be $P = p_L L + (1 + r)C$.

Thus, suppose the rate of interest is 5 percent, and that it takes, for the average manufacturing process, 500 beanbricks for fuel and 95 man-hours of labor to produce 1,000 beanbricks. Then the wage rate will be 5 beanbricks per hour, for, by equation (1): $(5 \times 95) + (1.05 \times 500) = 1,000$. Of the gross output of 1,000 beanbricks, 500 will go for capital depreciation, 25 for interest on capital, and 475 for wages.

In equilibrium, one additional relation must be satisfied. Many different processes, using different capital-labor ratios, are available to beanbrick manufacturers. They will use those processes that are most profitable at the current wage rates and beanbrick prices. Thus, if there existed a process, with input coefficients a'_1 and a'_2, such that the unit cost of production, $p_L a'_1 + (1 + r)a'_2$, was less than one, beanbrick manufacturers would find it profitable to expand their production using this process. In particular, at equilibrium, the rate at which beanbricks can be substituted for labor, or labor for beanbricks, by going to different available production processes, must be equal to the ratio of their prices—i.e., to $p_L/(1 + r)$. In the numerical example, 1.05 units of labor must be substitutable for 5 beanbricks in production.

the effect of technological change

Technological advance means improvement in some process, or the introduction of a new process so that more beanbricks can be produced than before with a given labor and capital input. The whole pie will be larger, and it remains to determine who will get the extra slice. The labor share per hour is simply the real wage rate (p_L in Equation 1), while the total income going to labor is the real wage times the amount of labor supplied. The latter quantity, the labor supply, equals the total output of beanbricks times the quantity of labor input required per beanbrick—the labor input coefficient, a_1, in Equation 1. Correspondingly, the total capital share is the rate of return, $(1 + r)$, times the capital input coefficient, a_2, times total production.

When technological change reduces the input coefficients, a_1 and/or a_2, Equation 1 no longer holds; the cost of production per beanbrick becomes less than unity. To reestablish equilibrium with the new input coefficients, one or more of the factor prices, the real wage rate or the rate of interest, must rise. The relative amount of the rise in each determines who benefits from the technological change, and by how much.

Now let us pay attention to one special case of technological change that can serve us as a yardstick and a point of reference and comparison for other possibilities. Let us consider a technological change that reduces the labor input per beanbrick, a_1, and leaves the capital input per beanbrick, a_2, and the rate of interest, r, unchanged. Then, we see from Equation 1, that in order to establish the new equilibrium, the real wage must rise. It must, in fact, rise in the same proportion as a_1 decreases, for the second term in the equation is, by assumption, unchanged. Labor, under these special conditions, will capture the entire increase in production resulting from the technological advance.

Notice that in the special case we are considering, the technological change is labor-saving, for the ratio of capital to labor in production, a_2/a_1, is increased. By the same token, we may call the change *capital-intensifying*. But sometimes we use *capital intensive* in a different sense to mean that the amount of capital per unit of output is increased. Our special case is not capital intensive in the latter sense, for we have assumed that the capital coefficient is unchanged.

Let us next consider what happens as a result of technological change that does not conform to the special assumptions we have been considering. If the capital coefficient increases while the labor coefficient decreases (the decrease more than compensating for the increase), then we can see from Equation 1 that, as long as the rate of interest does not increase, real wages must again rise to reestablish equilibrium, though not by a sufficient amount to capture the full gain in productivity. If, on the other hand, both labor and capital coefficients decrease, while the interest rate remains constant, then real wages will rise more than in proportion to the increase in labor productivity.

In sum, *so long as the rate of interest remains constant, an advance in technology can only produce a rising level of real wages.* The only route through which technological advance could lower real wages would be by increasing the capital coefficient (the added cost being compensated by a larger decline in the labor coefficient), thereby creating a scarcity of capital and pushing interest rates sharply upward. How likely is this chain of events?

The answer to this crucial question hinges on how capital intensive the technological change is likely to be—but capital intensive in the strong sense of requiring an increased capital input per unit of output. The fact of the matter is that most technological change—including mechanization or automation—is not capital intensive in this sense.

For example, while a modern computerized accounting department requires far more dollars of capital equipment per employee than it did a decade ago, it does not require as much capital equipment per unit of accounting work performed. A modern, highly automated power plant requires less capital equipment (in constant dollars) per KWH of energy produced than did the less automatic plant of a generation ago. A jet plane represents less capital per passenger mile than a 1920 biplane. We fail to note this decrease in capital intensity because we are accustomed to thinking of amount of capital in relation to labor employed, rather than in relation to output produced. As Kendrick and Sato show, however, the ratio of capital stock to output in the American economy decreased, in the forty years ending in 1960, at a rate of more than one percent per year. They calculate the ratio as 4.58 in 1919, and only 2.62 in 1960.[6] Mechanization and automation have been capital-saving as well as labor-saving, but because relatively more labor than capital is saved, we have been led to overlook this easily established fact. Under the assumption, which in the light of these facts seems valid, that technological change will lower, or at least not increase, the capital coefficient, can we say anything about what will happen to the rate of interest? Notice that if the capital coefficient remains the same, so does the ratio of capital stock to production for the economy. But modern theories of saving suggest that at a constant rate of interest, savings should increase proportionately with income, that is, in the right amount to produce the new capital required by technological change. Hence, we conclude that with a constant capital coefficient interest rates will also remain approximately constant. By the same token, with a declining capital coefficient, interest rates might also be expected to go down, for saving required to maintain the stock of production capital will now constitute a smaller and smaller fraction of income as production rises.

Again, our conclusions are not seriously inconsistent with the facts reported by Kendrick and Sato. For they find that although

---

[6] Kendrick and Sato, *Factor Prices*, Table I, col. 25. From Tables 4-1 and 4-2 in Kendrick, *Postwar Productivity*, it can be seen that this trend continued during the period 1960–66.

interest rates have not gone down in the past forty years, they have remained almost constant, and have certainly shown no strong tendency to rise.[7]

We conclude from our theoretical analysis that automation can only cause an increase in real wages, and, under realistic assumptions about the nature of the change and of the supply of capital, almost all the increased productivity will go to labor. Our confidence in this conclusion is considerably strengthened by its confirmation in the historical record. Kendrick and Sato show that the productivity of labor increased in the American economy over the forty years up to 1960 at a rate of about 2.35 percent per year, while real wages during this same period have increased at an annual rate of 2.56 percent.[8] As a result, the labor share in income rose from 72 percent in 1919 to 78 percent in 1960.[9] The earlier study covered gross private domestic product, including households and private nonprofit institutions; the later study, only private domestic business.[10]

the path to equilibrium

In our analysis to this point, we have simply compared the new equilibrium with the old, without indicating how the system would pass from the one to the other. To dispel doubts about the validity of the argument, it may be useful to trace the path of

[7] *Factor Prices,* Table I, col. 18. This conclusion is not refuted by the steep rise in interest rates since the late 1960s. This rise is due to anticipations of rising prices, hence represents a rise in the nominal (money) interest rate, and not in the rate adjusted for changes in the price level. If the annual rate of rise in prices is subtracted from the nominal interest rate, little or no increase remains in the net rate.

[8] *Ibid.,* Table I, cols. 20 and 17. According to Kendrick, *Postwar Productivity,* Table 4-3, the corresponding figures for 1948–1966 are 3.09 percent per year for labor productivity, and 3.30 percent per year for real wages.

[9] Slightly different figures, but the same trend, are shown by Kendrick, *Postwar Productivity,* Table 4-4: 67.3 percent in 1929, 75.3 percent in 1960, and 72.5 percent in 1966.

[10] Kendrick remarks, *Postwar Productivity,* p. 73, that "there was a drop in the labor share between 1960 and 1966, but the increase was subsequently resumed."

adjustment in some detail. Readers who are already persuaded that the conclusion is sound can dispense with this exercise.

Let us suppose, as before, that a technological improvement in some process, or the introduction of a new process, reduces the input coefficient to the point where the improved or new process is profitable at current prices. The new process will begin to replace existing processes.

Since the new process need not use the two factors of production in the same proportions as they are available in the economy, the supply of one or the other of these will presently be exhausted. Suppose that the new process is "labor-saving," which, in this context, means that it is more capital intensive, in the weak sense, than the economy as a whole has been. Then beanbrick capital will all be utilized while some labor is still unemployed. The effect of this will be to raise the price of beanbricks both relative to wages and absolutely. (The latter because, if not, high profit rates would lead manufacturers to bid up the price in their efforts to expand production.)

Notice that even a relatively small (percentage) saving in either the labor or capital requirements for production may yield an enormous return on capital at current wage rates. In our previous numerical example, we assumed a fuel requirement of 500 beanbricks for each 1,000 beanbricks produced, and a labor requirement of 95 man-hours. If the labor requirement were reduced by 10 man-hours (a little more than 10 percent), and if wages remained constant at 5 beanbricks per hour, the total return to capital, on a gross investment of 500 beanbricks, would rise from 525 beanbricks to 525 + 50 = 575 beanbricks — an increase in the net return per annum from 5 percent to 15 percent. Similarly, if the fuel requirement were reduced 10 percent, to 450 beanbricks, the gross return to capital would become 575 beanbricks, a net return of 17 percent per annum. Thus, we do not have to suppose that the supply of capital is very elastic in order to conclude that the requisite quantities of additional capital will be called out by these large opportunities for profit.

The short-run effect of a rise in the beanbrick price relative to the wage rate will be to make labor-intensive processes, using a relatively larger amount of labor and less beanbrick fuel, become more profitable. This will restore the balance of labor and capital demand to the ratio of availabilities of these factors in the economy.

But in the longer run, beanbrick capital prices cannot remain above their normal level of $(1 + r)$, for a higher price will divert large quantities of beanbricks from consumption to the capital supply. This will increase the beanbrick-labor availability ratio, will lower the beanbrick price, and make capital-intensive processes utilizing the additional beanbricks more profitable. Presently, a new equilibrium will be reached in which: (a) the ratio, on the average, in which labor and capital is used in the production processes is equal to the ratio of supply in the economy; (2) beanbrick fuel will sell at the normal price of $(1 + r)$; and (c) the use of new or improved processes will have reduced the average input coefficients, a_1 and a_2, to smaller values.

What will be the effect of this on real wages? Solving Equation 1 for real wages, we have:

(Equation 2) $\qquad p_L = (1 - (1 + r)a_2)/a_1.$

The second term in the numerator is always less than unity (the cost of beanbrick fuel is less than total output.) Hence, reducing a_2 will increase the numerator, and therefore increase real wages. Similarly, reducing a_1 reduces the denominator and increases wages. We conclude, as before, that *any technological improvement, regardless of whether it conserves capital or labor, will increase real wages.*

Let us again consider some simple numerical examples. Suppose a technological improvement reduces the fuel requirement from 500 to 400 beanbricks per 1,000 beanbricks produced, and the labor requirement from 95 man-hours to 10. The real wage before the change was 5 beanbricks per hour. The new real wage, from Equation 2, is $p_L = (1000 - (1.05)400)/10 = 58$. The wage rate has been increased by a factor of more than 10—somewhat more than the increase in direct labor productivity. The ratio of capital to labor has been increased by a factor of almost 10, but the ratio of capital to output, obviously, is unchanged.

As one further numerical example, suppose the labor requirement is reduced from 95 to 10 man-hours per 1,000 units of output, but that the capital requirement is increased from 500 to

600. As we have already shown, such an increase is unlikely, but it will provide an instructive example of an extreme case. The new real wage will be $p_L = (1000 - (1.05)600)/10 = 37$ beanbricks per hour. Now, before the technological change occurred, equilibrium in the economy called for a capital supply of a little more than 5 beanbricks for each man-hour of labor. After the change, the capital supply would have to be 60 beanbricks for each man-hour of labor. If the labor supply were constant, the stock of capital would have to multiply 12 times in the process of reaching the new equilibrium. However, the capital supplied, as a fraction of total output, would only need to increase by 20 percent. Such an increase could easily come about through a modest rise in the rate of interest.

restatement: technological change
and real wages

For the kind of economy under consideration, technological change, whether capital-saving or labor-saving or both, is bound to increase real wages. I have carried the argument through in detail—perhaps in somewhat tedious detail—because it is not found in any very prominent place in the modern economic literature (although it is neither a new nor an unorthodox argument), and because the practical issue is so important.

It is wise, however, to be suspicious of long chains of formal reasoning carried out within the confines of an abstract model. It is reckless to accept the conclusions of such arguments, much less apply them to questions of public policy, until it is reasonably certain that the conclusions do not depend in any critical way on the artificialities that were introduced into the model to make it tractable. The test of reasonableness can take two forms. First, we can test the model and our conclusions from it against the events of the real world. Does it fit the facts? Second, we can examine the model more closely to see exactly which are the critical assumptions for reaching the significant conclusions. We can then form judgments about the reasonableness of these critical assumptions.

Our main conclusion was that technological advance must, in the long run, raise real wages. A secondary conclusion was that if the technological change is labor-saving, it will increase the amount of capital per worker. We have already used the data of Kendrick and Sato and of Kendrick to test these conclusions against the recent history of the American economy. The first conclusion is also easy to test against the history of the Industrial Revolution in western Europe and America, and against more recent industrialization in other countries. Industrialization and increased mechanization have consistently and persistently been accompanied by rising real wages. The opposite relation appears only occasionally over short intervals of time and is submerged in the long-run trend. As we have seen, this finding is as true of recent events in the United States, in the face of automation, as it was in previous periods.

The second conclusion is harder to test, and we cannot consider all the technical issues here. Most technological innovations seem to call for a higher capital-labor ratio in production than prevailed before the invention. This does not necessarily imply (because of changes in the overall mix of production processes) that the capital-labor ratio will increase in the economy as a whole, but it makes such an increase plausible. In fact, there appears to have been a moderate, but relatively consistent, increase in the ratio of capital to labor in most countries during the period of their industrialization. More important, the increase in the ratio of capital to labor does not imply an increase in the ratio of capital to total income—and historically, this ratio has been declining in the United States.

The conclusions of the abstract model, therefore, fit the historical facts well. But what are the crucial assumptions that make the model behave as it does? The most important assumption is that the total amount of capital responds quickly and elastically to the price paid for capital, while the total amount of labor supplied responds only a little to the real wage. For, in equilibrium, the cost of production (including profits) must equal the price of the output. If production becomes more efficient, and the price of capital cannot rise, then the real wage rate must rise to balance out the equation. All of this can be read out of Equations 1 and 2. The rest of the analysis was needed only to show how the equilibrium is reached after a technological change occurs, and to make sure that we had the proper numbers of equations and variables to make the system determinate, but not overdetermined.

There is one other, less obvious assumption, mentioned earlier, to which attention should again be drawn—the assumption of constant returns to scale. This means that if the amount of labor and the amount of capital were doubled simultaneously, the total output would double. It does not mean that successive equal increments in labor input, with the quantity of capital held constant (or increases in capital, with labor held constant) would yield equal increments in output. On the contrary, the model assumes diminishing returns for capital and labor inputs separately, but constant returns for joint capital-labor inputs. This would be a dubious assumption in an agricultural economy that was pressing against the limits of its supply of good arable land. It is an assumption that may also require some modification for an economy that is pressing against other resource limits (say fuel limits).

What about the plausibility of the other main assumptions —the elasticity of the supply of capital and the inelasticity of the supply of labor? The essentiality of these assumptions to the argument is shown by the fact that David Ricardo and other eighteenth-century economists made almost exactly the opposite assumptions and reached almost exactly the opposite conclusions. They assumed that the supply of capital was relatively fixed. On the other hand, they assumed that if wages rose above the subsistence level, the population would increase to pull them down again (the "iron law of wages"). It was easy to deduce from this assumption, combined with the assumption of diminishing returns, a Malthusian equilibrium of population at the starvation level, and economics became the Gloomy Science.

However relevant the Malthusian assumption may be even today for much of the unindustrialized part of the world, it does not fit, and did not fit, the conditions of western Europe, the United States, or Japan. First, as we have already suggested, diminishing returns have not been an important consideration in most industrial economies. Second, the industrializing economies were generally able to accumulate capital a good deal faster than the labor force grew.

In considering the rate at which automating economies call for growth in capital, it is important to remember that although the increase required is more rapid than the increase in labor force, it is not more rapid than the increase in production. That is, mechanization and automation can be supported, with full employment, without any increase in the percentage of total income that is saved, but with that percentage constant or even declining. The required capital is generated automatically by the increase in production, and does not require a rise in interest rates to bring it forth.

Because of this capacity to generate the necessary capital, the industrializing economies have been able to absorb large amounts of technological progress that, on balance, has been labor-saving. That the rate of saving has responded appropriately to the forces of technological change (that is, has remained relatively steady in the face of rising incomes) is demonstrated by the stability of interest rates and the modest share of profits in total income during the periods of industrialization. With few exceptions, labor has received, in salary and wages, from half to three-quarters of the total product, and, in spite of increased capital intensity, this fraction has tended to grow rather than decline.

Finally, nothing about the current advances in automation indicates that these advances will have any different economic effects than earlier industrialization and mechanization. The main long-run effect of increasing productivity is to increase real wages—a conclusion that is historically true and analytically demonstrable. The speed of change does not appear to be greater than in previous eras. If a disequilibrium appears, it will evidence itself first in high profit levels, and the proper measures to restore equilibrium, if our analysis is right, are to encourage rapid private or public capital formation, which will come about at least partly because of the profits themselves.

A too rapid population increase will reduce the capacity of an economy to absorb capital-intensive technological change by slowing the growth of the capital-labor ratio. Too rapid wage increases may also depress the return on capital and slow the rate of new capital formation. In this case, disequilibrium can take the form of persistent unemployment. Judged from the events of recent years, however, neither of these dangers is a particularly threatening one for the American economy. Automation, and technological advance generally, are not the places to look for the causes of unemployment here.

Skilled and Unskilled

The formal argument showing that technological progress is fully consistent with full employment at gradually rising real wage rates is based on the assumption that labor is a single, uniform factor of production. We still must ask whether in the actual world, where the labor force represent a heterogeneous spectrum of skills, technological advance may place workers occupying particular skill levels at a disadvantage compared with others, or even render them unemployable.

Two diametrically opposite and contradictory concerns of this kind are expressed from time to time. In the fifties and sixties, it was sometimes asserted that the relatively unskilled part of the work force was unemployable in a high-technology economy—that only the educated, those with high school or college diplomas, would be able to find jobs. More recently, during the current depression, the opposite fear has been more often expressed—that the economy would be unable to employ the constantly increasing fraction of new entrants to the work force who held college degrees. Both of these concerns seem exaggerated. Let us consider them in turn.

employability of the unskilled

The evidence we reviewed in Chapter 2 indicates that automation does not, in fact, substantially change the distribution of skill levels in the factory or office. Moreover, among service occupations, the class of occupations most like to increase in relative numbers in the future, there are many opportunities for persons without special technical skills or advanced schooling.

A large fraction of the jobs in our present economy, perhaps as many as 70 percent, require no skills more complex than those involved in driving a car. But the latter skills are acquired by almost all of the adults in our society. Hence there seems to be no lack of jobs for persons with modest capabilities. When emergency demands are made upon the labor force, as in the Second World War, we find that most people have worked most of the time at skill levels below their best capabilities; and most people, appropriately motivated, are able to acquire impressive new skills, both military and civilian, in a great hurry and with modest amounts of formal training. The shipyards in wartime, or mechanized divisions moving across battlefields, provided dramatic evidence of the flexibility of the labor force.

It is true that unemployment has been highest among the most poorly educated and least skilled segments of the labor force. Employers, given a choice, have generally chosen educated over uneducated applicants and school graduates over school dropouts. But the preference is to be interpreted as a relative rather than an absolute preference. Nor is it clear that the criterion of choice has anything to do with skill. More likely, educational history serves as a surrogate measure of diligence and dependability.

Minimum wage laws and welfare allowances set a floor on the wages of the unskilled. This floor has always moved upward as general wage rates have moved upward, and more or less proportionately with them. We know little as to what happens, as we move toward higher technologies, to the *relative* productivities of skilled and unskilled labor. If the former increase more rapidly than the latter, the unskilled might be rendered unemployable at the accepted minimum wage. If this were to occur (there is almost no evidence, one way or the other, as to its likelihood), it would not invalidate our argument about rising average real wages, but it would require some institutional measures to be taken to maintain the opportunities of the unskilled for employment. The negative income tax proposal and schemes for guaranteed incomes are possible methods, already widely discussed, for dealing with this problem.

employability of college graduates

The current depression has aroused concern that the labor market may be unable to absorb a large fraction of the new entrants graduating from college. Again, the evidence we have examined does not indicate that automation is reducing the fraction of employed persons requiring the higher skills. What is happening, of course, is that an ever larger fraction of the American population is going to college. Without a fairly radical change in the occupational spectrum—of which there is no indication—many of these college-educated workers will not be able to get jobs of the kinds that persons with college degrees have traditionally sought and obtained.

However serious this problem may become, it is not profitable to treat it as a problem of technological unemployment. The kind of continual upward mobility that most of our population has experienced in the past is only possible in a society that is continually receiving new recruits at its lower levels—either through differential birth rates, immigration, or rural-urban migration. With or without technological change, there will have to come about in our society changes in expectations about mobility and about the accessibility of those occupations, like the professions, that are regarded as the most genteel, and as "appropriate" for educated persons.

This is not the place to speculate about how our society will adapt to the rising educational attainments of its labor force. To some extent, job requirements adjust to the availabilities of skills, just as skills adjust to job availabilities. To some extent, the meaning of a college education may change, and its vocational significance may decline. Whatever the modes of adaptation, there is little connection between this problem and the progress of technology.

The Information Technology and Resource Limitations

The economic argument that has been developed in this chapter leads to the conclusion that technological advance in general and automation in particular have not caused and will not cause the unemployment or impoverishment of human labor. Only brief attention has been paid above, however, to some of the serious problems that have been prominent on the public agenda during the past five or ten years: the possibility of exhausting resources, especially energy resources, and the equally threatening possibility of exhausting the capacity of earth, air, and water to absorb the polluting byproducts of production. Before we can be satisfied that full employment and continuing technological progress are genuinely possible into the indefinite future, we must be sure that these limitations will not stand in the way.

Because the issues are complex, we shall have to look at the problem from several points of view. First, we must consider the relation between technological progress and world population, for while the limitations that face us may be limitations on technology, they are also limitations on population. Second, we must reconsider what we mean by *technology*, and, in particular, disentangle the notion of technological advance from the notion of increasing or even constant capital intensiveness of production.

technology and population

In the past, the advance of technology has always been accompanied by a corresponding rise in the rate at which mineral and other resources are used and the volume of pollutants of all kinds that are produced. The magnitude of the resource and pollution problem that faces the world has grown with the product of world population times per capita energy consumption (or some comparable index of the level of technology). That product can be decreased either by decreasing population or by lowering the level of technology (defined in this way) Hence the problem of resource and pollution-absorption limits is completely interwoven with the population problem. The former cannot be solved in any long-run sense, whatever the level of technology, without solving the latter.

Of course population is related to technology in another way also. It was the advance of medicine, and especially public health measures, that reduced the Malthusian restraints on population growth to produce the great population explosion of the past century. Our humanity forbids our wanting to turn back the hands of that particular clock. If mankind were, collectively, more foresighted than it is, it would have taken pains to accompany the application of the medical technology with an equally widespread application of the technology of population control

—a technology that is available and, if used, fully adequate to the need. No such foresight, of course, was exercised, and we are now faced with the alternatives of making up for that lapse in the relatively near future—certainly within a generation or two—or letting nature handle the problem in its own effective, but highly painful, way.

I have no inclination to forecast how the problem will be solved, whether by the exercise of prudence or at the cost of immense misery. I would simply make three points about it. The first is that it has to be solved, independently of what we do about technology. The second point is that we possess the means to solve the health problem and the population problem simultaneously, that is, to maintain a stable world population without using infant mortality as the instrument of stabilization.

The third, and very important, point is that, at whatever level population is stabilized, only a society that possesses a high technology will be able to maintain standards of health, comfort, and well-being that we would regard as acceptable. The myth of the happy savage is just that—a myth. At present world population levels (and we are certainly unlikely to stabilize population at lower levels), adequate food production alone, ignoring all other needs, requires a high technology. But even with much smaller population numbers, we would be unwilling to return to a society in which at least half or three quarters of all labor was spent just in producing a subsistence level of food, clothing and housing. For that is what a society without high technology is like.

It is not a question of bringing the whole world's population, present or prospective, up to the American level of energy consumption. Human needs could probably be met at reasonable standards with an energy consumption of only one half or even one quarter of this amount. Even without further population growth—a totally unrealistic assumption—that would still mean more than a doubling of present rates of energy production, and a correspondingly larger strain on resources and the environment.

It would seem that we are ground between the upper millstone of resource limits and the lower millstone of human needs. In particular, there seems to be little place for new technologies that will make greater demands upon resources. Does this not mean that technology has to be stabilized along with population, and that technological advance must be halted? And must not automation and other advances in the new information technology be halted along with it? To answer these questions we must take a closer look at information technology, and particularly at the demands it makes upon resources and the environment.

technology is knowledge

Over the past several hundred years, there has been a close connection between technological progress and rising demands upon resources. Most fortunately, that connection does not hold for some of the important new technologies now being created. To see how this can be, we must understand what technological progress really means. Basically, technology is not "things," although some of it may be embodied in things. Basically, technology is knowledge: knowledge of how to make things, but also knowledge of how to do things. It is knowledge that broadens the range of alternatives before us. Our technology permits us to build large coal-fired generating plants that spew out, as a byproduct, oxides of sulfur and other noxious substances in their stack gasses; but advances in technology also tell us how to remove those substances and dispose of them safely. Advances in technology provide knowledge of new ways to pollute; but they also provide knowledge of ways to reduce pollution. They provide knowledge of new ways to produce and consume energy; but they also provide knowledge of new ways to conserve energy.

Today we are making important changes in our social accounting systems. We are insisting on taking a longer view of the costs of the resources we are consuming, and a broader view of the environmental and other nonmarket effects of our actions (what the economist calls *externalities*). An important goal of research and development at the present time—and certainly also in the future—is to find new alternatives for action that will conserve resources, make new resources available, and reduce or eliminate harmful side effects of productive activity. A related important goal of research is to develop techniques for analyzing the behavior of large, complex, interdependent systems, and techniques for monitoring such systems. We need such techniques if we are to plan our actions with due attention to their indirect environmental and social effects. Hence, continuing technological progress hinges, to an important and growing extent, upon advances in our information processing capabilities. We need to improve greatly our capacity for making effective managerial and public policy decisions.

The first phase of the Industrial Revolution was an energy revolution. The principal new technologies it provided were energy-producing and energy-using technologies. Today, that revolution continues, but mainly in the new directions of discovering ways to conserve energy, to produce it with less drain on scarce resources, and to eliminate deleterious effects of its production and use. For the rest, the Industrial Revolution is turning from energy-based technologies to information-based technologies—means for producing symbolic information and for processing it automatically. Because information-processing technologies are basically concerned with improving the processes of decision, we would expect them to become more and more important in a society that is oriented toward conservation, and that has to take into account an ever-widening, interrelated network of external consequences of its actions.

Resource and environmental limitations will not bar us from introducing improved techniques for information processing. To be sure, information processing is itself a consumer of modest quantities of energy and materials—mostly in the form of instruments and computers, and of the energy required to operate them. But we have been moving for thirty years toward producing ever larger quantities of information with a given input of glass, metal and energy. And as miniaturization of computer hardware continues, information-processing technology will be even more conservative of resources than it has been to date. Moreover, it is specifically upon the developments in information processing that we must rely for making best use of the energy and materials we consume in our society.

The concerns we have today, therefore, regarding the limits of resource availability and environmental capacity do not generally apply to information technology. On the contrary, the continued development of techniques for processing information will provide some of our principal new means for making the best use of available energy and materials.

Conclusion

In the first part of this chapter, it was shown that any level of technology and productivity is compatible with any level of employment, including full employment. In the last part, it was suggested that the problems we face today will not cause us to retreat from high technology—for such a retreat would not be consistent with meeting the needs of the world's population—but that they will bring about a substantial qualitative shift in the nature of our continuing technological progress. For future increases in human productivity, we will look more to the information-processing technologies than to the energy technologies. Because of resource limitations and because of shifting patterns of demand with rising real incomes, a larger fraction of the labor force than at present will be engaged in producing services, and a smaller fraction will be engaged in producing goods. But there is no reason to believe that we will experience satiety of either goods or services at full employment levels.

Technology is knowledge, and information-processing technology is knowledge of how to produce and use knowledge more effectively. Modern instruments—those, for example, that allow us to detect trace quantities of contaminants in air, water, and food—inform us about consequences of our actions of which we were previously ignorant. Computers applied to the modeling of our energy and environmental system trace out for us the indirect effects of actions taken in one part of our society upon other parts. Information-processing technology is causing all of us to take account of the consequences of our actions over spans of time and space that seldom concerned us in the past. It is placing on us—perhaps forcing on us—the responsibilities of protecting future generations as well as our own. In this way, the new technology, the new knowledge, is redefining the requirements of morality in human affairs.

Assuming new responsibilities is frequently painful. We see many difficulties in the world today that we did not see ten years ago. Sometimes we despair of dealing with all of the difficulties that confront us. We can take comfort, perhaps, in recognizing that there are really not more problems; there is just an increasing awareness of what the problems are. The information-processing technology is playing a major role both in producing that recognition and in providing new alternatives for handling the problems. The science of management decision, and the information-processing techniques on which it rests, will be important in determining whether we can discharge the broader responsibilities that we, as a society and as individuals, have accepted.

index